Wind Turbines

What Works
What Does Not
And Why

Saeed Quraeshi

Wind Turbines
What Works
What Does NOT
And Why

Copyright © 2017 by Saeed Quraeshi

Photos and Pictures licensed by - depositphotos.com
License ID #   10507152

ISBN -13   978 - 1548551971   (hard cover)

ISBN -10   154855197x   (flexi cover)

Printed by   CreateSpace, An Amazon.com Company

DEDICATION

This book is dedicated to my loving wife, Kari,
for all her support, encouragement and patience,
my children, grandchildren and
future generations.

ACKNOWLEDGMENT

I am grateful to all those who have supported me in my search for energy solutions over the years and indulged my passion for discussion.

My thanks to Martin Presse Booya for his help, and encouraging me to write this book.

I appreciate the special contributions made by Morgan Gelber and Alexia Quraeshi in the preparation of this book.

To Sabrina Quraeshi, my deep appreciations for her cover design and thoughtful suggestions.

My special thanks to Judy Steiner for her guidance and diligence in the editing, and proofreading of this book.

PREFACE

"The answer my friends is blowing in the wind"
Bob Dylan

Harnessing wind energy with wind turbines designed with the latest advancements in wind power technology would:

➤ Eliminate "NIMBY" (Not In My Back Yard) syndrome
➤ Be cost competitive with other energy resources
➤ Enhance market growth
➤ Improve sales
➤ Make wind power acceptable, reliable, and affordable

WHAT IS WIND ENERGY?

Wind is a renewable energy resource. Wind energy is an option that is infinite, renewable, sustainable, and environmentally friendly. The origin of wind energy conversion systems goes back to ancient times when it was realized that energy in the wind can be extracted by almost any device that can be made to oscillate or rotate in the wind stream.

Before the 1970s, wind energy, as a viable energy option for the generation of electricity was considered impossible. It was inconceivable to think that electricity generated by wind power would ever become cost competitive with electricity generated by power plants utilizing non-renewable conventional energy resources. In the year 2016, wind power was one of the leading cost competitive energy options for the generation of electrical power.

Prior to 1980, the design and development of wind turbines was based on building prototype units, and relying on trial and error to see if the design performed as expected. Attempts to scale up the size of wind turbines resulted in failures, mainly because of the lack of understanding of the dynamic behavior of wind turbines under operating conditions.

Since 1972, the advancements in computer technology and developments in computer based analytical tools, and design programs, and understanding the behavior of turbine blades during operation, significant improvements in the design of wind turbines have been made. Funding support for research and development (R&D) by governments in many nations helped the wind turbine industry to achieve success. This led to the emergence of the wind turbine industry. As a result, we now have wind turbines that are capable of generating electrical power that is cost competitive with electricity generated by conventional power plants utilizing non-renewable energy resources.

Rapid progress is taking place in energy storage technology to improve reliability. Drastic reductions in costs, due to mass production, are already proving that wind power will become affordable.

KEY MESSAGE

If we want to have affordable energy available to future generations, then all types of energy resources need to be utilized in a safe and clean manner.

To achieve this, we have to demand a shift in patterns of our energy consumption from utilizing conventional non-renewable energy resources to utilizing sustainable and environmentally safe renewable energy resources.

It is prudent of us to act wisely and make the shift, if not for our generation, then, as the legacy we leave to future generations to come. The future of wind energy depends on how the wind turbine industry implements innovations to design wind turbines that specifically address the issues such as "clients' concerns and demands" with respect to acceptability, reliability, and affordability.

Advancements in wind turbine technology now make it possible to eliminate the inherent design limitations of the existing wind turbines on the market.

Hybrid wind/solar systems will play a major role in supplying electrical power needs for all sectors of the economy.

WHAT YOU WILL FIND IN THIS BOOK

This book should be of particular interest to all those who are interested in learning about wind energy and its potential to generate electrical power that is cost competitive with electricity generated by conventional power plants utilizing non-renewable energy resources. Each chapter in this book is independent and as such can be used as a quick reference.

TABLE OF CONTENTS

CHAPTER 1

PERSPECTIVE ON ENERGY

1.1   OVERVIEW ON ENERGY

In the next decade, the availability of energy and the cost of energy are likely to remain the two most widespread issues facing both developing and developed nations.

The future of our energy-hungry world depends on our willingness to reduce our use of conventional non-renewable energy resources such as coal, oil, and nuclear fission. As our needs for energy resources become more complex, so must the ways in which we produce electrical power.

All available technologies, both non-renewable and renewable, must be explored for the generation of safe and clean power in order to provide maximum protection to consumers, taxpayers, and ratepayers.

We must ensure that we do not pass the burden of cleanup and waste disposal resulting from our use of conventional non-renewable energy resources such as coal, oil, and nuclear fission, as our legacy to future generations.

Renewable energy systems, such as wind power systems, must become cost competitive with conventional power systems, and prove to be acceptable, reliable, and affordable, by eliminating inherent limitations associated with their design.

Non-renewable energy such as coal, oil, gas, and nuclear power systems must become safe and clean, and remain cost competitive without any hidden costs.

It is only possible to achieve sustainable energy supply at an affordable cost when governments have the political will to commit to switch over to renewable energy resources and provide energy alternatives. All energy options for the generation of electrical power should be judged based on "Criteria for public concerns and demands". All energy options must be compared and selected based on being satisfactory to the public, namely:

➢ **Acceptable**
➢ **Reliable**
➢ **Affordable**

1.2   POPULATION GROWTH AND ITS IMPACT

There are many estimates of growth in world population. The following Table 1-1 illustrates "Global Population Growth" since the year 0 AD. The important point to note here is that the population of the world has grown at an alarming rate in the last century, and especially since 1950. The growth in world population from the year 1950 to 2017 illustrates the explosive rise in world population from two and one half (2.5) billion in 1950 to over seven (7.0) billion in the year 2017.

Table 1-1 Growth in world population

YEAR	Estimate of Global Population in millions
0 AD	150+
100 AD	200+
500 AD	300+
1000 AD	400+
1500 AD	500+
1900 AD	1,600+
1950 AD	2,500+
2000 AD	6,000+
2017 AD	7,000+
2050 AD	9,000+
2100 AD	11,000+

The rise in energy consumption is based on industrialization and increasing world population. Consuming non-renewable energy resources has resulted in causing severe air pollution and global warming. In addition, it has triggered the dramatic depletion and rise in price of all non-renewable conventional fuels, such as oil, coal, and nuclear.

1.3   POPULATIONS AND ENERGY CONSUMPTION

Skeptics always question if the goal of replacing non-renewable energy resources is ever achievable. By a close examination of the following Figure 1-1, one can appreciate the magnitude of the problem of population growth and demands on energy supplies.

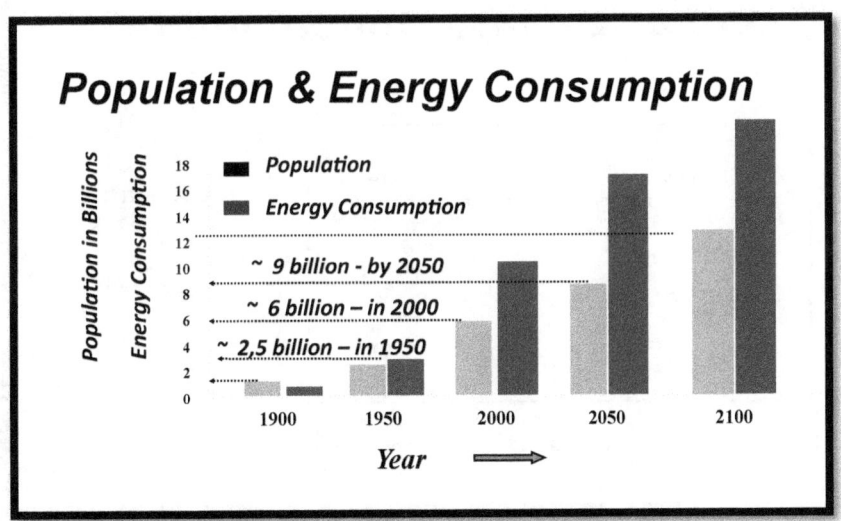

Fig. 1-1 Growth in world population

As a result, the demands on energy supplies and pressures for the development of new energy sources are increasing. We have to avert the catastrophic consequences of the shortages of energy supply due to:

> Dwindling supply of conventional non-renewable energy resources.
> Growth in global population.
> Increases in global demand for energy.

In the following Figure 1-2, statistics on global energy consumption in the year 2007 suggest that global renewable energy (solar, wind, hydro and others) consumption was about 10%. Nowadays, many countries are planning to increase the utilization of renewable energy resources. There is a growing consensus among many nations that it is feasible to achieve the goal of 50 % ~ 70 % renewable by the year 2050.

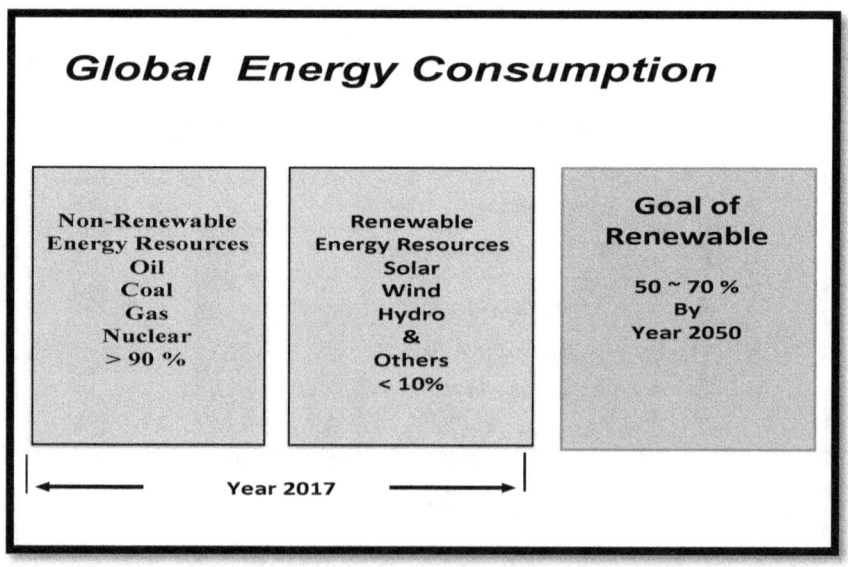

Fig. 1-2 Global energy consumption

To achieve a sustainable clean energy supply at an affordable cost, it is imperative that we demand the shift from conventional fossil fuel such as oil, coal, natural gas, and uranium, to renewable energy sources such as wind, photovoltaic, solar thermal and biomass; energy resources that are infinite, renewable, sustainable and environmentally friendly.

Unlike a renewable resource, the costs and economics of power generation by conventional resources will be governed by the costs for the fuel. Presently, the development of the electricity supply sector is characterized by a dynamically growing renewable energy market and an increasing share of renewable electricity.

This shift will compensate for the phasing out of plants utilizing conventional non-renewable energy resources. If urgent actions are taken, then by 2050, about 70% of the electricity produced worldwide could come from renewable energy sources, such as wind and solar.

The real constraints on the deployment of renewable energy technologies lie not in the abilities of these technologies to meet the challenges of product development, but rather in the course of action to be undertaken by the governments of the nations involved.

Since the 1970s the potential of utilizing renewable sources of energy for the future are gaining serious consideration. If and when the governments and the electrical power utilities will have the political will to commit to switch over to renewable energy resources and provide energy alternatives; it is only then possible to achieve a sustainable energy supply at an affordable cost.

1.4   GROWTH IN ENERGY CONSUMPTION

Historical records on "Global Energy Consumption" show that prior to 1880, forest products, water power, and coal were the main sources of energy. During the period from 1880 to 1920, hydro, natural gas, coal and wind resources were utilized to generate electricity. The use of oil was introduced in 1920. In 1960, nuclear energy was introduced as the most promising source for electrical power generation.

The energy crisis in 1973 resulted in uncertainties regarding the future prices of non-renewable conventional energy resources. Later in the 1970s and early 1980s, accidents in nuclear power plants and the high costs of building them traumatized the world into reassessing the nuclear option. It also made us aware that as promising as it may sound, the advances in science and technology can have grave consequences if issues related to safety and pollution are ignored.

This has led us to a better understanding of the earth's ecosystems, and the need to assess the impact on society and the environment. It has also led us to search for the availability of risk-free, environmentally friendly, and sustainable forms of energy resources for electrical power generation. Since 1956, as the global demand for electricity grew, due to global industrial expansion and the unprecedented growth in world population, conventional power plants using non-renewable energy resources such as coal, oil, and nuclear were utilized to satisfy the global demands for electrical power generation.

If we are to avert the catastrophic consequences of the shortages of energy supply, then it is imperative that we demand the shift from conventional fossil fuel such as oil, coal, natural gas, and uranium, to clean and sustainable renewable energy sources such as wind, solar, solar thermal, and biomass. Global energy consumption refers to the amount of total energy used by all human beings worldwide.

There have been many authoritative attempts to estimate our future global energy demands. Several of these studies have taken the year 2000 and/or 2030 as their time horizon targets. However while some may regard these studies as speculative, others may try to disregard the conclusions.

Global energy consumption has increased steadily for much of the twentieth century, particularly since 1950. Over the last century most of our energy has come from conventional non-renewable sources such as coal, oil, natural gas and nuclear.

A review of the world's renewable and non-renewable energy resources indicates that the depletion of non-renewable resources is a matter of time, while the renewable resources provide us with a hope for a better future.

At the current rate of depletion of our limited conventional resources, we should get more serious about the utilization of renewable resources for electrical power generation, which will be vital to our energy future.

In order to resolve this energy dilemma, all nations must address the same basic questions. How can one:

➢ Increase the reliability of energy?
➢ Supply and ensure the gradual replacement of hydrocarbons?
➢ Utilize the natural energy resources within one's own national boundaries to provide better protection of consumers?

> ➤ Provide acceptable forms of energy to reduce global warming?
> ➤ Preserve the environment from pollution?

1.5   ENERGY OPTIONS

Non-Renewable Energy Resources: The main non-renewable energy resources utilized in the generation of electrical power can be divided into two main groups:

1. Fossil fuels   (Coal, Oil, and Gas)
2. Nuclear fuels

Coal: Coal is a non-renewable resource. It is compressed organic matter, plants and vegetation, which has been deposited deep between layers of sediments within the earth and entombed for millions of years. Coal is reclaimed through mining. Coal is an ideal energy source; as it tends to be highly combustible, and when ignited produces a large amount of energy.

Oil: Oil is a non-renewable resource that builds up in liquid form deep in the earth. Oil is extracted by drilling deep into the ground and pumping the liquid out. The liquid is then refined and used to create many different products such as gasoline, diesel, heating oil, propane jet fuel, and plastics.

Gas: Natural gases are a non-renewable resource. They are buried below the earth's surface and, like oil, must be drilled for and pumped out. Ethane and methane are the two common types of gases obtained through drilling.

Nuclear   Nuclear fuel is also a non-renewable resource that is used to produce energy. It is obtained through mining and refining of uranium ore.

Fossil fuels are non-renewable because they will run out one day. Burning fossil fuels generates greenhouse gases and relying on them for energy generation is unsustainable.

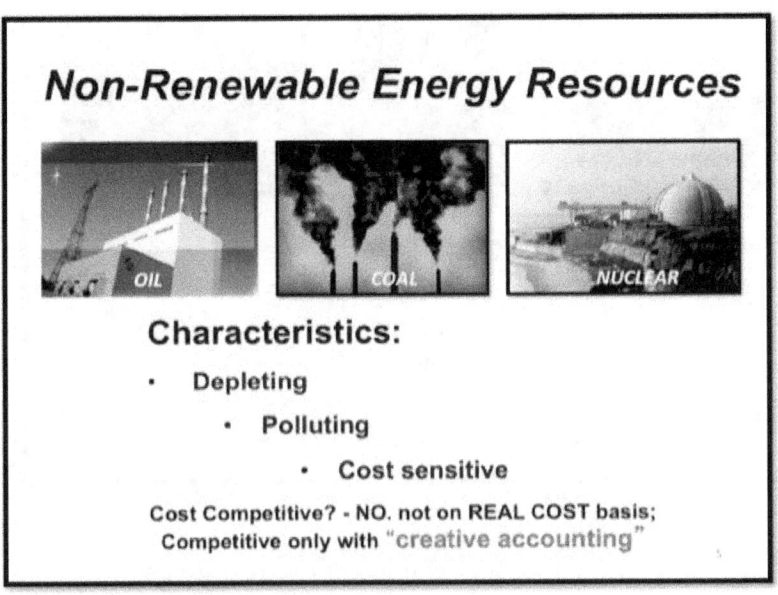

Fig. 1-3 Characteristics of non-renewable energy resources

Renewable Energy Resources   Electricity   generated   by renewables played a critical role, having accounted for around 90% of new electricity generation in 2015. The leading renewable energy resources utilized in the generation of electrical power can be divided into three main groups namely:

> ➢ Hydro
> ➢ Wind
> ➢ Solar

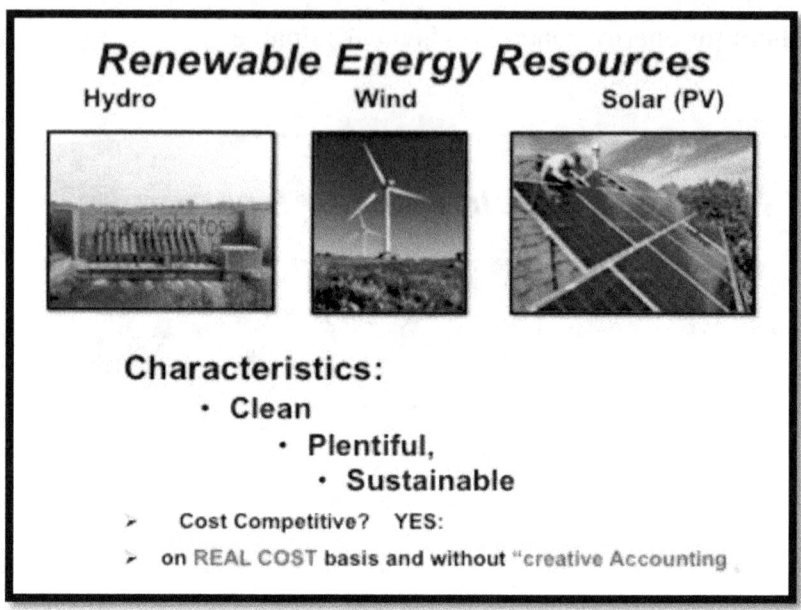

Fig. 1-4 Renewable energy resources

Renewable energy resources provide maximum protection of consumers, the taxpayers, and ratepayers, but also ensure that we do not pass the burden of cleanup and waste disposal as our legacy to future generations.

1.6   SUMMARY

There has been an explosive rise in world population, from two and one half (2.5) billion in 1950 to over seven (7.3) billion in 2017, and it is estimated to reach nine (9) billion by 2050.

In the next few decades, the availability of energy and the cost of energy are likely to remain the two most widespread issues facing both developing and developed nations. Energy is an essential part of our daily lives. Even a minor interruption in its supply causes a major disruption of our social and economic lives.

Fossil fuels are non-renewable because they will run out one day. Burning fossil fuels generates greenhouse gases and relying on them for energy generation is unsustainable.

We must utilize all available technologies, both non-renewable and renewable, for the generation of power in a safe and clean manner, not only to provide maximum protection of consumers, taxpayers, and ratepayers, but also to ensure that we do not pass the burden of cleanup and waste disposal as our legacy to future generations. Use of renewable energy resources will:

✓ Achieve improvements in "Standard of Living" .
✓ Provide unlimited renewable energy resources.
✓ Curb the rise in cost of energy.
✓ Reduce depletion of non-renewable energy resources.
✓ Significantly reduce the generation of greenhouse gases.

If urgent actions are taken, then by 2050, more than 70% of the electricity produced worldwide could come from renewable energy sources, such as wind, hydro, and solar (photovoltaic).

CHAPTER 2

EARLY HISTORY OF WIND POWER

2.1   HISTORICAL OVERVIEW
(200 BC ~ 1850 AD)

Energy in the wind has been used for centuries for agriculture, transportation, and industrial purposes. Since 200 BC until 1980, for many applications, in locations with adequate wind conditions, both vertical-axis and horizontal-axis windmills were reliable, practical and cost competitive with other forms of energy.

The technology for windmills was developed when people realized that almost any device that can be made to oscillate or rotate in the wind stream could obtain mechanical power.

The earliest known vertical-axis windmill design dates from about 200 BC. Vertical axis windmills were developed in Persia, presently known as Iran, and used to grind grain and pump water. Records also indicate the development and utilization of vertical axis windmills in China.

Early immigrants brought windmills for water pumping application to America. Wind pumps were used to pump water from farm wells for cattle. An estimated 60,000 wind pumps are still in use in the United States.

Since the 12th century and up to the 1850s, the industrial revolution was solely powered by renewable energy, such as windmills.

Very large windmills, with rotors up to 18 meters (60 feet) in diameter, were utilized to supply water for the steam-powered railroad trains used for commercial transportation.

2.2   ANCIENT VERTICAL-AXIS WIND MILLS

The ancient Persian Windmill was a straight bladed vertical axis wind turbine, which consisted of:

> ➤ A bundle of reeds, or sailcloth, tied together, acting as blades. The blades were connected to a central shaft, which turned stones to grind grain. The end result was that the grain was ground into flour.

> ➤ A turbine located inside a housing structure was specially designed to ensure that wind would rotate the turbine in the desired direction.

> ➤ The entrance wall was shaped to provide augmentation of wind entering the turbine.

> ➤ The walls at the inlet of the vertical axis windmill were designed to increase the wind flow on one side to turn the rotor. It also protected the other half of the rotor by obstructing the incoming wind from slowing the side of the rotor that was moving toward the wind, thus reducing the drag on the rotor.

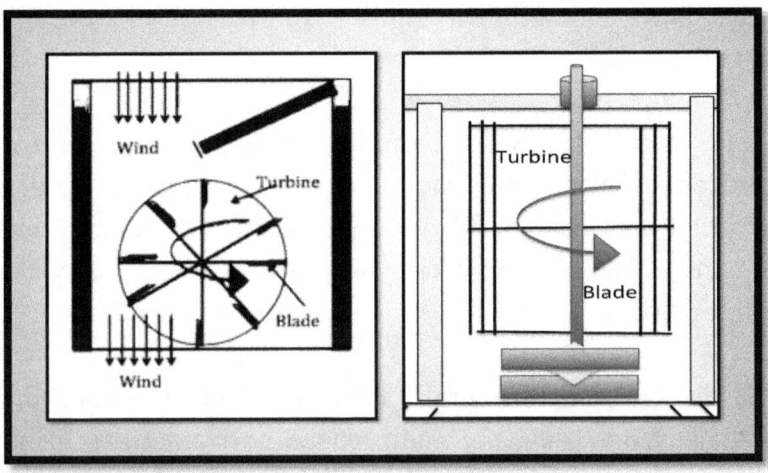

Fig. 2-1 Ancient vertical axis windmill

A typical Persian windmill had a turbine of 5m in diameter x 6 m in height, and the housing structure dimensions were about 6 m x 6 m x 10 m in height. Some of the larger windmills had larger turbines and were of 15-20 meters in height. They fulfilled their purpose and proved their design made them acceptable, reliable, and affordable.

Analysing the designs of these ancient vertical axis windmills indicates that:

➢ The windmills were of rugged construction for reliability.
➢ The windmills had outer walls shaped to provide augmentation of wind, which increased the air mass flow and wind speed, and made them suitable for operation in low wind speed conditions.

2.3   HORIZONTAL-AXIS WINDMILLS

The horizontal-axis windmills were invented in eastern Persia around 500-900 A.D.

The horizontal-axis windmills arrived in Europe through Morocco and Spain, and later on were brought to the north of the European continent by German crusaders. The rotating arms of the windmills, commonly known as sails, were normally covered with sailcloth.

Horizontal–axis windmills were installed in Europe in 1270 A.D. Later, by 1400, the multi-bladed horizontal-axis windmills that were developed and used were of a technologically advanced design with a gear train based on designs used on waterwheels in Europe.

It took another 500 years to perfect the design of horizontal-axis windmills to achieve a higher efficiency and superior performance. Windmills were used for a variety purposes, such as:

- ➢   Pumping water.
- ➢   Raising water to help drain wet areas to reclaim land.
- ➢   Extracting oil from grains and nuts.
- ➢   Making paper from old clothing.
- ➢   Making coloured powder for preparing dyes.

Wind and water, as primary industrial energy sources, played an important part in the industrial revolution in Europe. Windmills in Holland served many purposes. The most important probably was pumping water out of the wetlands, and back into the rivers outside the dikes, so that the land could be cultivated for agricultural purposes. At their peak, the total number of wind-powered mills in Europe is estimated to have been around 200,000. In 1850, there were about 10,000 windmills in use in the Netherlands, of which about 1000 are still standing.

Fig. 2-2 Ancient horizontal axis windmills

Eventually, after 1850, windmills were replaced by the development of steam engines and internal combustion engines. There was a rapid decline of wind energy as the primary industrial energy source.

At present, some of these horizontal-axis windmills have been preserved for their historic value.

Windmills were unable to compete with cheaper, more reliable and seemingly abundant sources of non-renewable energy such as coal, oil, and natural gas, for the following reasons:

> ➢   Small power producing capability of the horizontal-axis windmills ranging up to a maximum of 80 kW.
> ➢   Erratic operation due to the inherent nature (intermittent availability) of wind.

2.4   HORIZONTAL-AXIS WIND PUMPS

Early immigrants brought windmills such as wind pumps for water pumping applications to America. They were used to pump water from farm wells for cattle.

In the United States, the windmill was part of a domestic water system. For centuries, the most important use of windmills has been mechanical water pumping using small systems with rotor diameters of less than 10 m (32 feet).

These small windmills were used on farms for water supply for both the farm home and stock watering. The multi-bladed wind pump was a well-known fixture of the landscape throughout rural America.

These windmills for water pumping had a large number of blades, which turned slowly with considerable torque in low winds, and were self-regulating in high winds.

They consisted of a rotor connected to a gearbox and crankshaft, which converted the rotary motion into reciprocating strokes carried downward through a rod to the pump cylinder below.

Fig. 2-3 Windmill for water pumping

In the United States, in 1854, Daniel Halladay invented the self-regulating farm wind pump. Since than, the designs of the wind-pump were significantly improved. Between 1850 and 1970, over six million mostly small (1 horsepower or less) mechanical output wind machines were installed in the U.S. alone. An estimated 60,000 wind pumps are still in use in the United States

Wind pumps are used worldwide. A typical wind pump of about 5 m (16 feet) in diameter can start to operate in a wind speed as low as 1.5 m/s (3 mph). In a 6 to 9 m/s (15 to 20 mph) wind condition, it can pump about 6000 liters (1600 US gallons) of water per hour to a height of 30 m (100 feet).

Very large windmills, with rotors up to 18 meters (60 feet) in diameter, were used to supply water for the steam-powered railroad trains used for commercial transportation.

2.5 SUMMARY

When it comes to the use of renewable energy resources such as wind, we have so much to learn from the past. Analysing the designs of these ancient windmills indicates that:

✓ The windmills were of rugged construction for reliability. Windmills were acceptable to the public.

✓ They fulfilled their purpose and proved their design made them "acceptable", "reliable", and "affordable".

✓ Since the 12th century and up to the 1850s, the industrial revolution was solely powered by renewable energy, such as the horizontal-axis windmill.

✓ Without these windmills the industrial revolution in Europe would have been impossible. Steam engines and internal combustion engines replaced windmills.

✓ Since the early 1900s, windmills were unable to compete with cheaper, more reliable and seemingly abundant sources of non-renewable energy such as coal, oil, and natural gas.

CHAPTER 3

WIND POWER
BASICS

3.1   OVERVIEW

The only dependable energy that we can rely on in the future is that which we receive from the sun. The energy we receive from the sun is everlasting, abundant, and sustainable. That makes solar energy the most suitable for our power requirements.

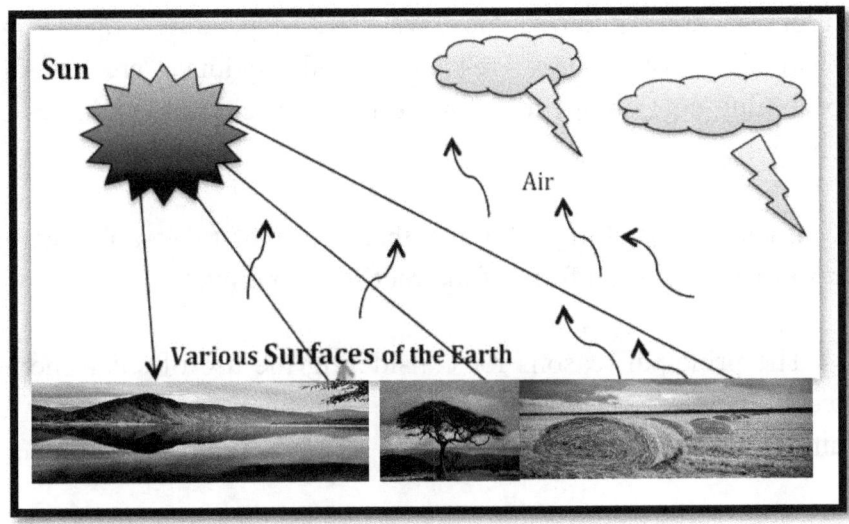

Fig. 3-1 Generation of wind by the sun

Air in contact with the earth's heated surface rises making low pressure in the area, and the air from the surrounding higher-pressure areas moves into the low-pressure area, thus creating the flow of air, known as wind. Movement of air is labeled as "wind". The faster the air moves the stronger the wind gets. Wind speeds can vary from a gentle breeze to a tornado.

The advancement in wind power and photovoltaic technologies since 1974 to the present date can be directly attributed to research, development, installations, testing, and gaining operating experience.

Experience to date demonstrates that wind and solar systems can be tailored to suit application modes, without constraints on the physical size of the system.

Utilization of commercially mature products, such as wind turbines and photovoltaic systems, depending upon the availability of energy resources and options, are rapidly becoming cost competitive with other non-renewable sources of energy.

Basically speaking, like all other forms of renewable energy resources, wind is a form of indirect solar energy.

The principal reasons for considering the use of solar energy technologies, for example wind and photovoltaic, are its attractive characteristics such as:

> Non-depletion of resources.
> Reduction of global warming and atmospheric pollution.
> Ease of conversion to electrical power.
> Zero fuel cost.
> Compatibility with winter and summer energy demands.
> Resources within one's own national boundaries.

3.2   VARIATIONS IN WIND SPEED

The knowledge of the various types of variation in wind speed is very important in planning the application and utilization of wind energy resources available at a specific site.

The flow of wind is highly influenced by the earth's surface; factors include such variations in topography as: vegetation cover, desserts, grasslands, trees, forests, valleys, mountains, buildings, rivers, lakes, and oceans.

The rotation of the earth also contributes to the direction of the planetary winds. Seasonal variations in the energy received from the sun affects the strength and the direction of the wind.

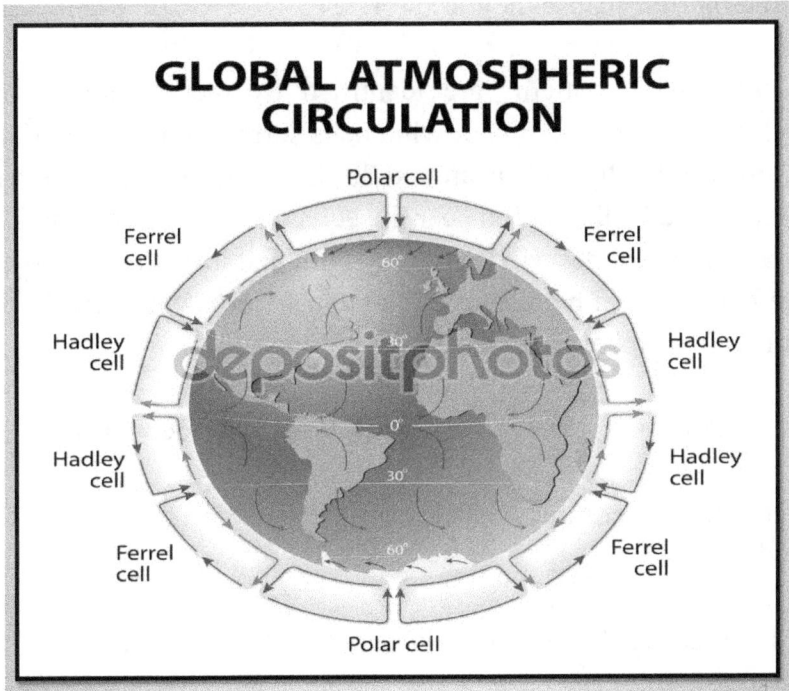

Fig. 3-2 Global wind patterns

Unequal heating of the surface of the Earth causes large global wind patterns. The uneven heating of the earth's surface by the sun generates planetary winds. The earth's surface at the equator is heated to a greater degree than the surfaces at the North Pole and the South Pole.

At any given time and place the wind may be substantially different from the established norms of the global pattern, due to the local topography and weather systems.

At the equator warm air rises and moves toward the poles. The cooler air at the poles sinks and moves back toward the equator. Nevertheless, global winds do not move from north to south or south to north because the Earth rotates.

Global wind patterns are identified by the direction from which winds blow. Six major wind belts surround the globe. Each of these wind belts' size is around 30 degrees of latitude. At zero degrees latitude, the region near the equator is called doldrums. It is a region where the winds are very light and irregular. There are three belts in each hemisphere. The upper three wind belts in the northern hemisphere are called:

(1) Polar easterlies, (2) Westerlies, and (3) Trade winds

And, the lower three wind belts in the southern hemisphere are called:

(4) Trade winds, (5) Westerlies, and (6) Polar easterlies

Winds in the Northern Hemisphere seem to curve to the right as they move, while in the southern hemisphere, winds appear to curve to the left.

This is known as the Coriolis effect, due to the rotation of the Earth. There are three distinct types of patterns in movements of air masses caused by the solar radiation reaching the earth and the earth's rotation.

> Diurnal Pattern - Different temperatures cause this at day and at night. This effect is more distinct at coastal sites than offshore.

> Depressions and Anti-cyclones Pattern- This pattern is more distinct in oceanic than continental regions. It typically occurs in cycles of about 4 days.

> Annual Pattern - This pattern varies with the degree of latitude and vanishes in close proximity to the equator.

These patterns are important not only for energy yield estimations, but also for forecasting of wind power output.

3.3   WIND VARIES WITH HEIGHT

The movement of air, generally recognized as wind, is greatly affected by the type of the earth's surface the wind is blowing over.

The speed at which the wind moves nearest to the surface of the earth is much slower than the speed at higher elevation.

The wind near the surface of the earth is slower due to friction with the earth's surface.

Due to the retarding effect on the wind speed by the earth's surface, the wind speed increases with the height above ground at any given location.

According to the wind shear formula, the speed varies as a power of the height. One of the standard formulae for speed v as a function of height z is:

$$v(z) = v\,10\,(\,z\,/\,10\,\text{m}\,)\,\alpha$$

Where v 10 is the speed at 10 m, and a typical value of the exponent α is 0.143 or 1/7. This phenomenon in recognized as "wind shear".

The following table indicates the impact of the earth's surface on the percentage of variation of wind speed with respect to the height of the wind above the earth's surface. The greater the height the stronger the wind speed gets. In open areas, the wind speed increases by 12 % every time the height above the ground is doubled.

Table 3-1 Variation of the wind speed

Height above surface (m)	Variation in Wind Speed
5	90 %
10	100 %
20	112 %
40	125 %
50	129 %
80	140 %
100	145 %

The device to measure the wind speed is known as an "anemometer". To obtain accurate wind speed conditions at a specific site, it is essential to use an anemometer.

Mean wind speed, the wind shear, and the probability distribution are important for analysis of wind turbine performance.

The details of overall measurements of wind speed over any given area in a country is available from the local "National Climate Centers" (NCC) or the "Atmospheric Environmental Services" (AES).

The measurements represent, in general, the wind pattern for a given locality. "Wind Maps" are prepared, which detail the wind speed measurements in various areas in the country. The following Table 3-2 details the various classifications of wind speed.

Table 3-2 Wind power class vs. power density

Wind Speed @10m ~ [33 ft.]		
Wind Power Class	Wind Power Density (W/m²)	Speed m/s (mph)
1	0	0
2	100	4.4 (9.8)
3	150	5.1 (11. 5)
4	200	5.6 (12.5)
5	250	6.0 (13.4)
6	300	6.4 (14.3)
7	400	7.0 (15.7)
7	1000	9.4 (21.1)

NOTE. Each wind power class should span two power densities. For example, Wind Power Class = 3 represents the Wind Power Density range between 150 and 200 W/m². The speeds are average wind speeds measured over a period of one year.

These parameters, together with density, describe the magnitude of wind power at any given site and, given the operating characteristics of a wind turbine, defines the amount of energy that can be extracted over a given period of time.

3.4   ANNUAL AVERAGE WIND SPEED

The overall wind conditions on an annual basis, on a specific site or in a region, are defined as "Annual Average Wind Speed" (AAWS).

It is the wind that would flow over a period of a year, (i.e. 365 days x 24 hours = 8760 hours/year), if all the variations in the wind flows over the year were uniformly rounded out.

Generally speaking, the most likely wind speed is estimated at about 75% of the average wind speed. Normally, wind speeds of more than two times the average wind speed do take place occasionally, but not very often, usually a total of less that 200 hours/year.

3.5   ENERGY AND POWER FROM WIND

When the radiant energy from the sun, commonly known as "sunshine", reaches the earth, it is converted into mechanical energy of the moving air masses. Mechanical energy is the sum of potential energy and kinetic energy.

Depending on wind speeds, the raw kinetic energy can vary from a gentle breeze to a hurricane with awesome force. The rate of flow of kinetic energy in the wind is called wind power. The faster the movement of wind over a location on the earth's surface the more power there is in the winds. The total mechanical power available in a given wind stream is equal to the volumetric flow rate times the kinetic energy per unit of volume of that wind stream.

The formula for measuring the available power in the wind flow is expressed as:

Available Power = ½ ρ A $V^{3.}$
Where Available Power = Power in watts (W).
ρ = air density (1.225 kg/m^3 at standard conditions).
A= area of the wind stream (m^2).
V= wind speed or velocity (m/s).

Note that in the formula, the power of the wind is proportional to the cube of the wind speed. This represents the cube law relationship between wind speed and wind power.

It means that the wind speeds higher than the average possess much more power. If the wind speed doubles, the kinetic power in the wind will increase by eight.

Table 3-3 Variation in wind speed vs. power

Wind Speed in (m/sec)	Wind Power in Watts (W)
2	8 ~ (2x2x2 = 8)
3	27 ~ (3x3x3 = 27)
4	64 ~ (4x4x4 = 64)
5	125 ~ (5x5x5 =125)
6	216 ~ (6x6x6 = 216)
7	343 ~ (7x7x7 = 343)

For example, the cube of two (2) is eight (8 =2 x 2 x 2). It means that doubling the wind speed means an eight-fold increase in power. Maximum power that any wind turbine can convert is equal to 16/27 or 0.593 times the power available in the wind, that is - maximum power = $0.593 \times \frac{1}{2} \rho$ A V.

An ideal wind turbine with 100% aerodynamic efficiency would attain an output equal to this maximum power or 59.3% of the available power. The factor 0.593, commonly known as the Betz's limit, therefore corresponds to the maximum theoretical value of the power coefficient. In practice though, output between 30% and 50% of the available power can be extracted with well-designed wind turbines. The size of wind turbine needed to power a home depends on a home's energy use, annual average wind speeds at the location, and the turbine's height above ground. Generally speaking, a typical American home would require a small turbine with a 5 kW generating capacity (Average 8000 ~ 9000 kWh/year) to meet all its electricity needs.

3.6   UTILIZATION OF ENERGY RESOURCES

The following Figure 3-3 indicates the various factors that influence the costs associated with utilization of energy resources.

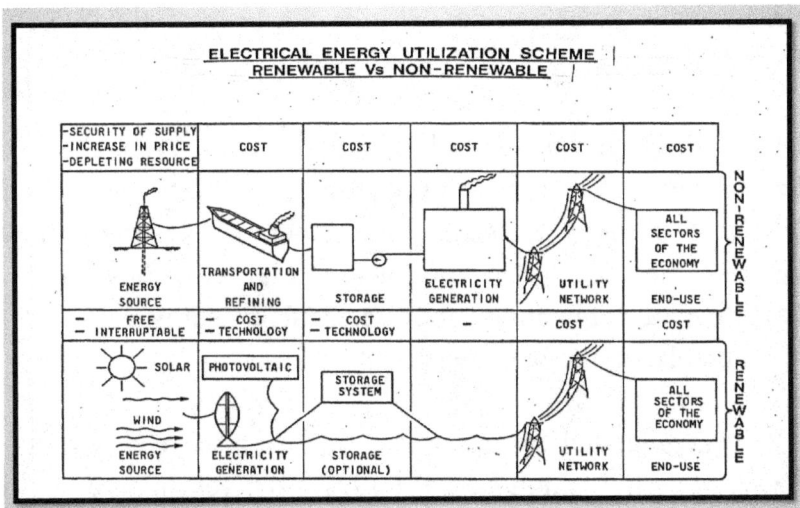

Fig. 3-3 Energy utilization scheme

It illustrates the utilization of the two energy forms, renewable and non-renewable. The basic difference between the two systems is in the location of the energy storage system.

For an on-demand power supply, due to its characteristics of being an intermittent source of supply, renewable energy resources require energy storage after the generation of power.

The energy utilization scheme, given in Figure 3-3, clearly illustrates that energy storage is essential for an on-demand power supply.

For conventional non-renewable energy resources energy storage is required before (prior to) the generation of power. In some cases the storage may be optional, depending on the type of generation mix available in the utility system. Short, medium, and long-term storage can be achieved by a variety of storage concepts such as supplement to hydro, batteries, flywheels, and production of hydrogen by electrolysis. In general, for a short storage cycle time, batteries provide the lowest initial cost and life span cost.

Ever since the energy crisis in the early 1970s, there is a renewed interest in promoting the possibility of utilizing renewable energy resources to generate electrical power. However, electrical power utilities have always considered any alternative to conventional non-renewable energy resources as unacceptable, on the basis that:

Power generated by renewable energy resources is not cost effective due to its high capital cost. By the very nature of being an intermittent source of energy, power generation by renewable energy resources is not suited for an on demand power supply. It requires energy storage to provide an on demand power supply.

Earlier energy storage systems utilizing batteries were either in the demonstration stage, unproven in service, cost intensive, or technically not suited to meet electrical utility standards.

Since the year 2000, drastic reduction in costs, (over 80 %) have been achieved. Such cost reductions can be attributed to:

> Mass production.
> Significant advancements in technology, and in product development.

Since 2005, the cost of energy storage systems has decreased from around 1200 US$/kWh to about 250 US$/kWh in 2016. It is estimated that by the end of the next decade, the cost of battery storage systems would be reduced to between 150 and 100 US$/kWh.

Since the year 2010, worldwide, there are more than a dozen companies that are leading in the development of batteries. These companies are heavily investing in the energy storage market.

At present, battery systems are in production to suit the requirements of all sectors of the economy, namely: residential, commercial, industrial, electrical utility, and transportation.

Successful applications have demonstrated battery storage systems to be safe and reliable. Battery storage systems are now being deployed on many projects to provide cost effective energy storage for solar and wind energy systems.

This will guarantee the acceptance of renewable energy systems as well suited for an on demand power supply.

There are several sources of energy to generate electrical power such as: potential, chemical, thermal, kinetic, and nuclear energy. These can be further categorized as non-renewable and renewable.

There are several advanced solar energy conversion systems in the developmental stages.

Technological advances in energy storage systems would certainly enhance the utilization of wind and solar technologies in the energy sectors.

3.7   SUMMARY

✓ The energy we receive from the sun is everlasting, abundant, and sustainable. The only dependable energy that we can rely on in the future is that which we receive from the sun.

✓ Basically speaking, like all other forms of renewable energy resources, wind is a form of indirect solar energy. All the energy required for the masses of air to flow across the surface of the earth comes from the sun.

✓ In areas unrestricted by variation in the surface of the earth, wind speed increases by twelve percent (12%) for every time the height above the surface is doubled.

✓ The overall wind conditions on an annual basis, on a specific site or in a region, are defined as "Annual Average Wind Speed". (AAWS)

✓ Normally, wind speeds of more than two times the average do take place occasionally, but not very often, usually a total of less that 200 hours/year.

✓ In general, for a short storage cycle time, batteries provide the lowest initial cost and the lowest life span cost. Latest technological advances in energy storage systems utilizing batteries are making rapid inroads in the energy storage market.

✓ This would certainly improve the use of wind and solar technologies in the generation of electrical power.

CHAPTER 4

LARGE
MW WIND TURBINE
TECHNOLOGY

4.1   OVERVIEW

Large multi-megawatt wind turbines are specifically designed for the generation of electricity for the utility grid system. Arrays of large wind turbines are commonly known as wind farms.

In 1941, the first MW-class horizontal axis wind turbine was installed on the mountain known as Grandpa's Knob in Castleton, Vermont, USA.

The unit was rated at 1.25 MW. It was named Smith-Putnam wind turbine. The unit was synchronized to a local utility grid in Vermont.

It operated for 1,100 hours before suffering a critical failure. The unit was not repaired, because of a shortage of materials during the Second World War. It would be the largest wind turbine ever built until 1979.

After the Second World War, the electrical power utilities emphasized the development of power plants based on relatively cheaper non-renewable energy resources, such as coal, oil, natural gas, and later nuclear power with a promise of an abundance of cheaper electrical power supplies.

It was only after the energy crisis in the early 1970s, when oil prices increased from 2 US$ /barrel to nearly 18 US$/barrel, that the interest in the development of renewable energy resources was renewed worldwide.

The wind turbines, we see today in wind farms in many countries of the world, are the result of successful commercialization of the wind turbine industry.

4.2   WIND POWER INDUSTRY

At present, both the Horizontal-Axis (HA) and the Vertical-Axis (VA) MW units are cost competitive in domestic and international markets for wind farms. HAWTs are currently generating the vast majority of wind power in today's market. The wind power industry has matured around HAWTs.

In addition, since 1990 wind farms are applying the best available technology to design HAWTs with improved performance in order to produce cost effective units.

The following Figure 4-1 represents examples of HAWTs installed on a wind farm.

Fig. 4-1 Examples of wind farms with HAWTs

A wind farm is a group of wind turbines located at a site used to produce electricity. A large wind farm may consist of several hundred individual wind turbines. The term "offshore wind power" is used to describe a wind farm constructed offshore.

The rationale to build offshore wind farms is: higher wind speeds are available offshore, energy production in terms of electricity generated is higher, opposition to construction, by the public, is usually much weaker, and the cost of offshore wind power has decreased drastically in the last decade making offshore wind farms cost-competitive.

The wind power industry is involved with the design, manufacture, construction, and maintenance of wind turbines. The large MW wind turbine industry began a program of testing prototype units and series production in 1979.

In response to the increase in oil prices in the early 1970s, in 1975, NASA managed a program for the United States Department of Energy (DOE) to develop large utility-scale wind turbines for electric power. Based on historical records of land based large wind turbines in USA and in Canada, on evolution of commercial wind turbines during the period between the years 1980 and 2008, indicates the following:

- ➤ Large MW VAWT technology in Canada - Project EOLE – 4 MW VAWT, 1980 ~ 1988. The development of wind energy program in Canada was cancelled in the mid 1980s.
- ➤ Large MW HAWT technology in USA - 1980 ~ 2008. In the mid 1980s, funding for further development of the wind energy program in the USA was curtailed due to budget cuts, and the wind energy program was cancelled. This led to wind turbine installations based on HAWT units imported from Europe and other countries.

The development of the commercial wind turbine industry in the U.S.A. was however delayed by a significant decrease in competing energy prices during the 1980s, and due to several failures of the wind turbines developed and tested under this pioneering program.

All of the earlier prototype models of multi-megawatt HAWT units, both in Europe and in the U.S.A., experienced serious operational problems within a year or less, and were removed from service.

Even though the government sponsored US and Canadian wind energy programs were terminated in the 1980s, the work on design and development of large utility scale wind turbines continued to flourish in European countries.

Despite the fact that VAWTs had a very successful prototype unit, unlike HAWT units, funding for further development was curtailed due to budget cuts, and the wind energy program in Canada for large MW units was cancelled in the mid 1980s. The horizontal axis wind turbines (HAWT) that we see today in the wind farms in many countries of the world are the result of successful commercialization of the wind turbine industry.

Although wind turbines are becoming cost competitive in the generation of electricity, with conventional power plants, the current wind power technology of both Horizontal-Axis and Vertical-Axis MW units, due to their inherent design limitations, are experiencing resistance from the public including the growing rise of "Not In My Back Yard" (NIMBY) syndrome.

The main two areas of problems with the existing design of the MW size HAWTs are:

> Reliability of the wind turbines.
> Acceptability of the wind turbines by the public.

For these two reasons the existing wind turbine industry is:

➢ Unable to meet the existing demand.
➢ Powerless to capture the growing market for wind turbines.

As the 21st century began, rising concerns over energy security, global warming, and eventual fossil fuel depletion led to an expansion of interest in wind energy.

Wind turbine production has expanded to many countries. Worldwide, HAWT units are now operating, with a total nameplate capacity of over 200,000 MW. Size and power ratings of large MW commercial wind turbines have increased greatly, with the Enercon E-126 capable of delivering up to 7 MW.

Currently, the MW wind turbine industry is experiencing a time of mergers and re-organizations. These companies are making substantial investments in the field of development and manufacturing, to meet the growing demand worldwide for large MW wind turbines.

In order to overcome the "NIMBY" syndrome, the wind turbine industry needs to overcome the "inherent design limitations" of large MW wind turbines. A paradigm shift in its approach to designing wind turbines is essential to achieve the goal of making the large MW units acceptable to the public.

4.3   DESIGN AND SYSTEM COMPONENTS OF HAWTs

Generation of electrical power is the main purpose of the wind turbine. The wind turbine rotor drives a large shaft, which connects to a gearbox, which in turn increases the revolutions per minute to a speed suitable for the electrical generator.

Wind turbines have either induction or permanent-magnet generators, or direct-drive, depending on the type of wind turbine drive-train system design. The wind turbine has a number of sensors for controlling the functions to monitor, such as: speed and direction of the wind, rotor speed, pitch angle of the blades, vibration levels, temperature of system components and several other variables.

An onboard computer processes the inputs to manage the operation of the turbine, and provides a safety system, which can override the controller in an emergency.

The control system protects the turbine from operating in dangerous conditions and ensures that the power being supplied to the electrical grid system has the proper voltage, current, and frequency.

The major system components of the HA wind turbine are:

- Base
- Support tower
- Blades
- Rotor
- Hub
- Nacelle
- Gear box
- Generator
- Controller
- Drive shafts
- Yaw system

However, there are over 8,000 components in each turbine assembly. The following provides a brief description of the system components in a HAWT:

Base   The base, generally made of concrete and steel, supports the wind turbine structure. The tower is fitted with a base flange, which is normally attached to the foundation by anchor rods embedded into concrete and bolted to the implanted tower stub. For the support tower foundation, depending on the condition of the ground at the site, where the turbine is being installed, various designs of concrete slabs, with multi-pile and mono-pile solutions have been used.

Support Tower   The support tower houses the electrical wiring and conduits, and supports the nacelle. It provides access to the nacelle for maintenance. The nacelle is mounted on top of a tall tower to allow the blades to take advantage of the higher wind speeds at an elevated level. For large MW wind turbines, the support tower elevates the turbine rotor to the desired height in the range of 50 to 150 meters.

Blades   Blades have an aerodynamic profile, like an airplane wing. Wind turbine blades use lift to capture the wind's energy, by spinning a generator in the nacelle. In 2015, the length of blades for a large multi-megawatt onshore wind turbine was 50 m (164 ft) and larger. The length of each of the blades for an offshore turbine was 80 m+ (262 ft+). The blades spin at a slow rate of about 10 to 25 revolutions per minute (RPM), although the speed at the blade tip can be as high as 150 to 200 miles per hour, depending on the size of the rotor. The size of the blades may range from about 30 to 150 meters. Normally, the blades are made of materials that have high strength-to-weight ratios, such as carbon fiber, and fiberglass.

Rotor   The rotor for a typical utility-scale MW wind turbine consists of three blades, a hub, and a low speed shaft. The blades with an aerodynamic profile generate lift, which makes the turbine rotor turn. They are bolted onto the hub, with a pitch control mechanism to permit the blade to rotate about its axis, and to help limit the rotational speed of the rotor under varying wind speeds.

Nacelle   The nacelle houses a generator and gearbox, and other essential drive train system components. The gears increase the low rotational speed of the blades to the generator speed. Rotation of the generator produces electricity, which is fed onto the electrical grid system. Certain wind turbines use a direct drive system. A direct drive system connects the rotor directly to a generator.

Yaw System   HAWTs need to constantly align themselves into the wind using a yaw-adjustment mechanism. All horizontal-axis wind turbines have a yaw drive system to keep the rotor facing into the wind. They also help to unwind the cables that travel down to the base of the tower. The yaw-drive system consists of gears and either an electrical or hydraulic motor mounted on the nacelle, which drives a pinion gear to stabilize the turbine during its normal operation.

Installation   Installations at a project site involve transportation of turbine components by road, rail, and water. Due to the increasing size, weight, and length of the system components, managing the heavy-load and long haul requirements is a daunting task. This involves moving blades that are over 30 to 60 meters (98 to 196 feet) long and weigh over 68,000 kg (150,000 pounds). Once the system components such as the nacelle, tower sections and blades are delivered to the installation site, construction of the wind turbines can start.

4.4   ADVANTAGES OF WIND TURBINES

There are many advantages of wind energy, to name a few:

➢ Wind is a renewable energy resource. It is environmentally friendly, as compared to conventional non-renewable energy resources. There is no pollution generated by the system's operation, no need to mine for fuel resources such as coal or uranium, or to drill for exploitation of oil, or natural gas, or to transport the fuel to burn and pollute the atmosphere.

➢ With proper wind turbine design, very low maintenance cost. After all costs for the installation are fully paid, then the power is essentially free, except for a minor cost for maintenance. Land near wind turbines can still be used for agricultural or other purposes.

➢ In most cases, wind/solar hybrid systems are the most cost-effective renewable energy option. Costs continue to decrease.

➢ Worldwide, there is an abundant domestic supply with enormous potential. The worldwide potential of wind power is more than 400 TW (terawatts).

➢ It is a sustainable energy resource, for another 6+ billion years, as long as the sun keeps shining.

4.5 DISADVANTAGES OF WIND TURBINES

There are also many disadvantages of wind energy, to name a few: Wind is an intermittent source of energy. Until a major breakthrough in energy storage technologies occurs, we will have to use wind turbines together with other conventional energy sources to meet our energy demand for reliability of energy supply.

Wind is highly variable. Not all places are suitable for wind energy. Wind speed varies by season, weather, and location. Therefore, evaluating projected wind-system output is difficult due to variability of turbine design and production conditions.

Existing wind turbines on the market face many obstacles, including the existing State or local municipal zoning laws, which may result in expensive hearings or could possibly prevent installation. Wind is random and the availability of wind energy is uncertain. Therefore it would require adequate energy storage for it to function as a base load system. Even though designs of wind turbines show significant improvements, there still remain a number of major issues with existing limitations in the inherent design of the HAWTs.

The major inherent limitations of the design of HAWTs are:

➢ Safety in operation.
➢ Birds and bats killed by the turbines.
➢ Noise generated by the rotation of the turbine blades.
➢ Damage and fire caused by lightning strikes on the turbine.

Safety in Operation   Safety in operation is a major concern to the public. For example:

➢ Blade throw during turbine operation.
➢ Ice throw during turbine operation, in winter.
➢ Damage or total destruction of wind turbine during operation, especially under over speed conditions.

Safety in operation is a major issue with all existing wind turbines, both large and small. Since the early 1970s, there have been numerous reports of accidents related to wind turbines. For example:

In 2015, a wind turbine at the Screggagh wind farm in Northern Ireland collapsed. Debris from the turbine scattered over hundreds of metres. There were no high windy conditions or bad weather conditions reported. The cause of the accident was not well defined at the time the accident occurred.

Even though there are thousands of wind turbines in operation, such accidents with wind turbines are possible, even with the latest models of wind turbines on the market.

Birds and Bats Killed   Birds and bats killed by wind turbines are a common occurrence.

In 2016, a proposed project that would have generated electricity from wind energy in southwestern Saskatchewan was denied over concerns about birds. Environment Minister Scott Moe said an environmental review of the proposal for Chaplin identified potential risks to migratory bird activity.

Algonquin Power wanted to build a 177- megawatt facility on behalf of SaskPower that would have included a maximum of 79 wind turbines, 50 to 70 kilometers of access roads and 110 kilometers of trenched transmission lines.

A review of a variety of reports on issues related to "birds and bats killed" indicates that US wind farms kill 10 to 20 times more birds and bats than previously reported. America's wind farms are actually slaughtering millions of birds and bats annually. It is projected that by 2030, the United States plans to produce about

20% of its electricity from wind. That's nearly six times as much as in 2016. By 2030, wind turbines in the USA would be killing over 3 million birds and 5 million bats annually.

In 2012 study conducted in Spain concluded that Spain's 18,000 wind turbines are killing 6-18 million birds and bats yearly.

No country in the world can afford to ignore the killing of the birds and the bats by wind turbines, nor its impact on the ecology, agriculture, and economy. If this killing of birds (raptors) continues then it will cause rodent populations to soar. If the killing of bats continues then it will have a profound impact on agriculture.

Bats also control forest pests and serve as pollinators. Swedish studies have documented their attraction from nine miles away to insects that swarm around wind turbines, hence the slaughter.

With proper design such accidents with wind turbines can be eliminated.

Problems of Lightning Strikes In most modern wind turbines, both small and large, the preferred material for blade construction is fiberglass, in order to make blades lighter in weight and to reduce stress. However, blades made of fiberglass and or composite materials are highly susceptible to damage by lightning.

Lightning damage to the turbine blades accounts for:
- Poor operational performance of a wind turbine.
- The highest failure rate of any single component.
- Warranty on the units is limited. After each case of severe weather condition, inspection of the turbine blades is

recommended, (in accordance with operation and maintenance manuals) which results in extensive downtime and additional cost for maintenance.

➤ Blade inspection and repair programs are one of the most common causes of wind turbine downtime in the wind energy industry. Financial impact of a lightning strike to a turbine blade is: costly property damage, and business interruption during inspection, repairs, and replacement.

➤ Almost 20% to 30% of insurance claims reported in the wind turbine industry.

➤ On large wind turbines, blade replacement could cost about $80,000 or $120,000, contingent on the extent of the damage to the blade.

Problem of Noise   Blade high tip speed is the main cause of the noise emitted by the turbines. The higher the rotating speed the louder the noise level.

The blades of small wind turbines rotate at an average range of 175-500 revolutions per minute. Large MW turbines' blades rotate in the range of 15-50 rpm.

Depending on the size of the turbine rotor and the rpm, blade tip speed may range from 190 mph to over 300 mph.

To achieve any reduction in noise created by the blades, the rotational speed must be reduced, which in turn would result in a drastic reduction in the power rating of the turbine, making the turbines rotor size and costs increase and therefore substantially increase the cost of energy production.

4.6   DESIGN OF LARGE MULTI-MW VAWT

Since then, there have been successful demonstrations of prototype VAWT units, ranging from 30 to 500 kW rating, in many countries including the USA, the UK, Netherlands, Sweden, Germany, and China, and with units in the range of 50 to 4000 kW in Canada. Vertical axis wind turbines have been the focus of the Canadian Energy program since its early interest in the technology in 1965. The following Figure 4-2 shows a picture of the 4 MW VAWT, located in Cap Chat, Quebec, Canada.

Fig. 4-2   Project EOLE - 4 MW VAWT

In the USA, over 500 first-generation VAWTs have been in operation for over ten years and have logged more than 10,000,000 operating hours, thus demonstrating the reliability and maturity of the design.

The VAWT technology has matured, based on the successful operation of wind farms in California with unit ratings ranging from 50 to 300 kW. Advanced second-generation VAWTs were competitive with commercially available HAWTs.

Earlier versions of VAWT unit installations during the 1970s and 1980s cover a wide range of sizes and kW ratings, [6,30,50,150,230,300,500 & 4000 kW].

Development and the successful operating experience with the 4 MW project EOLE in the 1980s added many technical innovations and improvements to VAWT technology It surpassed all commercially available HAWTs, in reliability, in operation, and performance.

Up until early 1990, HAWT units were designed to operate at constant speed and with a blade-pitch-control system to limit excess power output at high wind speed.

By comparison, all of the VAWT units were designed to operate at constant speed and with a fixed-blade-pitch system to control power output. This gave an edge to the HAWT in better performance over the VAWT unit.

4.7   VAWT TECHNOLOGY (1980 ~1990)

VAWT technology is a proven technology. VAWT technology is based on its merits of technical superiority and cost competitiveness.

Development and the successful operating experience with project EOLE in the 1980s added many technical innovations and improvements to the VAWT technology, such as:

> ➤ A direct drive generator with an AC/DC/AC asynchronous link for variable speed operation for improved performance, more energy capture, and utility grade power production.

These gave VAWTs an edge over HAWTs in superior performance. Based on the successful demonstration of superior performance of EOLE, HAWTs began to incorporate these improvements in their designs.

Unfortunately, funding for MW VAWTs ceased, which impacted further development of VAWT technology.

VAWT   Vertical axis wind turbine (VAWT) has been the focus of the Canadian Energy Council for over twenty years.

Canadian studies on comparisons of Large Wind Turbine Generators for Electrical Power Generation and operating experience gained over the five-year period (1988-1992) on the project EOLE (4-MW VAWT) confirm the viability of competitiveness of the VAWT in the wind turbine market.

Experience in Prototype Operation   During the 1980s, the multi-megawatt VAWT unit had a far superior track record in experience on operation.

> ➤ The very first multi-megawatt VAWT unit (4-MW Project EOLE) operated for over five years in continuous service.

> ➤ The vertical axis wind turbine has several advantages over its horizontal counterpart. Maintenance is far more practical, because all critical components are located at ground level.

> ➤ The project EOLE (4-MW VAWT) was the first large MW turbine to incorporate a direct-drive generator system with variable speed power train.

> ➤ The system has proven the technology with five years of successful operating experience on the world's largest VAWT unit. (4 MW project EOLE).

> ➤ It also was the first MW unit to ensure high quality power generation that was acceptable for utility grid supply requirements.

The following Figure 4-3 shows the configuration and the overall dimensions of the 4 MW VAWT unit known as project EOLE

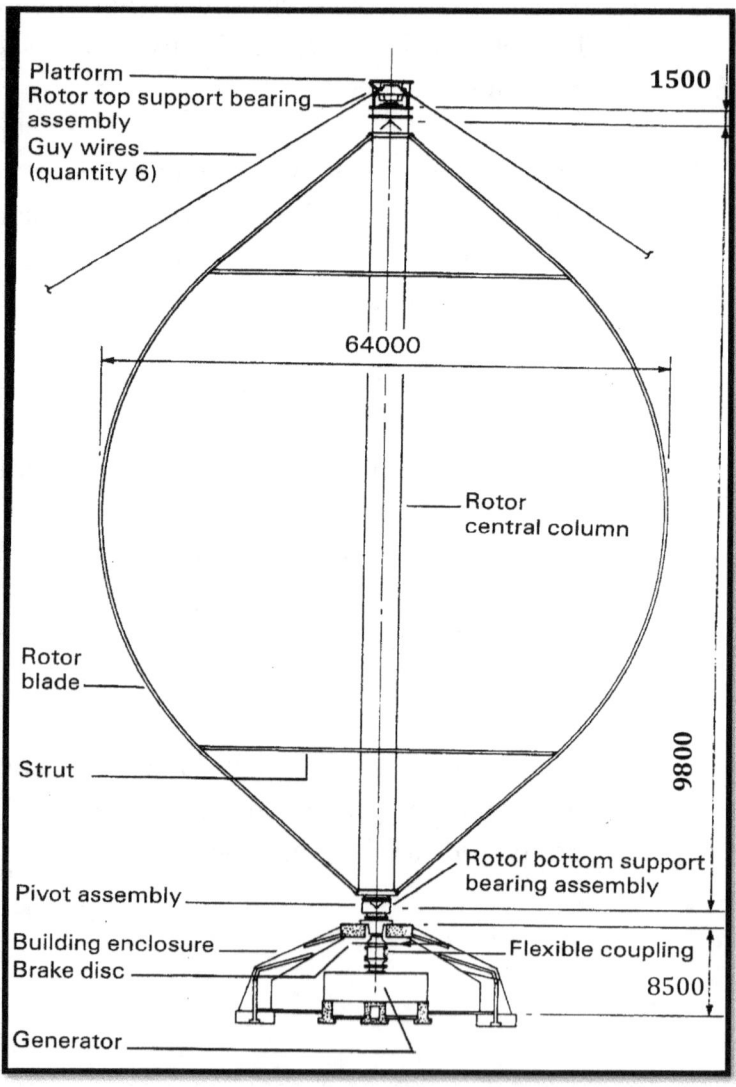

Fig 4-3   Project EOLE - 4 MW VAWT

4.8   HISTORY OF PROJECT EOLE

The following provides additional information on the operational history of Project EOLE.

During the entire period of over five-years of operation, Project EOLE encountered three problems, namely:

> ➤ Turbine operation at reduced speed.
> ➤ High stress levels in one of the blade-to-strut connections.
> ➤ Failure of the lower bearing.

Turbine Operation at Reduced Speed   After the installation of the unit, and during the initial start-up and commissioning stage, a management decision was taken to operate the unit at a reduced rotational speed, in a constant-speed mode. (About 12 rpm instead of the 3.6 MW rated power at 14.25 rpm. This reduced the rated electrical power output from 3.6 MW to about 2.5 MW)

Damper   This decision was taken mainly to avoid the lengthy testing period required to properly tune the "Damper Assembly" installed in the turbine column, the purpose of which was to eliminate harmful resonance(s) during variable speed operation of the unit. The fine-tuning of the damper would have required extensive testing. It was decided to postpone the proper tuning and commissioning of the damper system, until after the end of the 5-year R&D test period. The damper arrangement was dictated by vortex shedding of the column, and to damp out out-of-plane and in-plane modes. The damper was installed to provide the ability to operate the unit in a variable speed mode throughout the operational speed range, without resonance, and thereby extend the life of the machine.

High Stress Level   The two-bladed turbine has two blade support struts per blade, a total of four blade-to-strut connections. During the early period of operation, high stress levels were observed only in one of the lower "blade-to-strut connections". This problem was detected, and the blade-to-strut connection was repaired. After the repair, the unit operated without any problem.

Lower Rotor Thrust Bearing   The lower bearing failure problem occurred after the bearing had been in operation for over five years. The bearing manufactured by SKF was a prototype unit, and the world's largest bearing of its type.

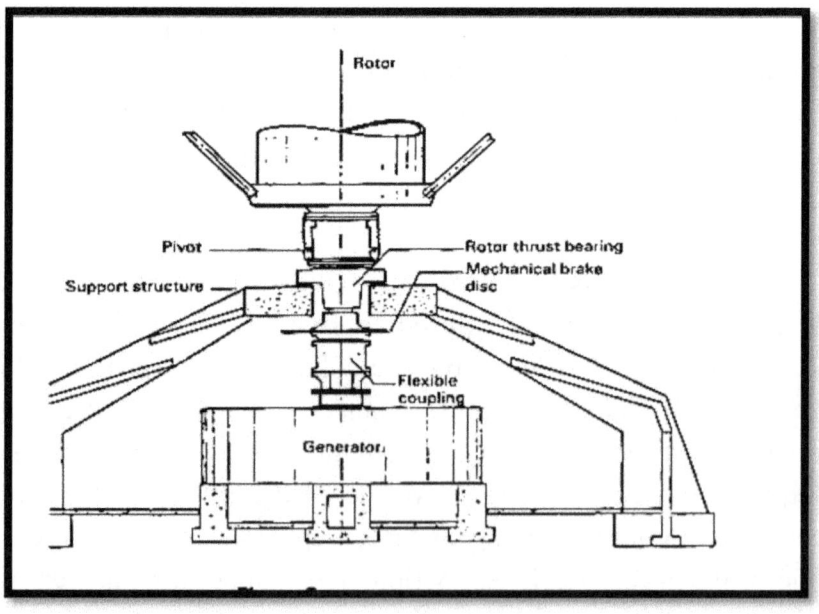

Fig. 4-4 Location of lower rotor thrust bearing

The total design load on the bearing, due to the weight of the turbine rotor and the thrust of the support guy cable system, was in the order of 400,000 Kg. (882,000 lbs.)

The bearing being a prototype, and a one of a kind, there was no guarantee, and the expected life of the bearing was not predicted. The probability of a bearing failure is that one bearing in a batch of 1000 bearings may fail prematurely, due to a variety of reasons.

To monitor the bearing in operation, sensors were provided to indicate its status. During the earlier years in operation, the bearing sensors tripped and stopped the unit several times, which indicated the start of a possible problem.

However, the sensors were disconnected, and the unit was kept in operation without monitoring the status of the lower bearing. The bearing failure occurred after nearly five years in operation. For the record, it must be noted that all commercially available bearings can fail and some do fail. Bearing manufacturers suggest that, based on MTBF (mean-time-between-failure) analysis, any bearing can fail. Therefore, the failure of a special, one-of-a-kind, prototype bearing, after a period of over 5 years in operation, after ample warning by the sensors of a possible problem, is not a failure of the design or concept.

The unit was shut down, awaiting a decision to repair the bearing. The bearing assembly is designed to be easily removable for repair and maintenance, without having to disassemble the whole turbine unit

To remove the bearing, the turbine is supported on a frame and jacked up, the pivot is removed, and the bearing can then be removed for repairs, if required.

Except for these three problems, the unit operated for over five years in continuous service with a 93.7 % overall availability. The unit was operationally unattended and remotely monitored to observe the daily performance of the project. It supplied utility quality power into the Hydro Quebec electrical grid. This record is a testimony to the success of the 4 MW VAWT Project EOLE design.

Prototype Experience in Operation   Designed   with FMEA and FTA analyses techniques, the Multi-megawatt VAWT unit has a far superior track record in experience on operation.

The very first multi-megawatt VAWT unit (4-MW Project EOLE) operated for nearly five years in continuous service. EOLE was one of the world's largest, longest running, most reliable megawatt-scale wind turbines – either VAWT or HAWT. Project EOLE Operation Record is as follows:

> Commissioned   March 1988
> Grid Coupled   18,550 hours
> Starts   Over 6,000
> Energy Output   12,000 MWH
> Availability   94 %
> Shut down   April 1993

Many of the earlier models of multi-megawatt HAWT units, both in Europe and in the U.S.A., experienced serious operational problems within a year or less, and were removed from service.

The Multi-megawatt VAWT unit has a far superior track record in experience on operation. This record is a testimony to the superiority of the VAWT design. In 1993, due to a problem with the prototype bottom support bearing, which happened after five years in operation, the unit was shut down. The bearing was repairable, without having to take the whole unit down. However, the economic consideration of repairing and of keeping a single prototype unit operating was the reason that the unit was not put back in operation.

4.9   VAWT TECHNOLOGY (1990 ~ 2015)

Since the early 1990s, companies in many countries have promoted R&D work for VAWT technology. Advances made in the technology are carried out to keep VAWT technology alive and cost competitive and maintain the comparative lead position it had achieved in the 1980s. The following Figure 4-5 compares the competitiveness of the HAWT and the VAWT.

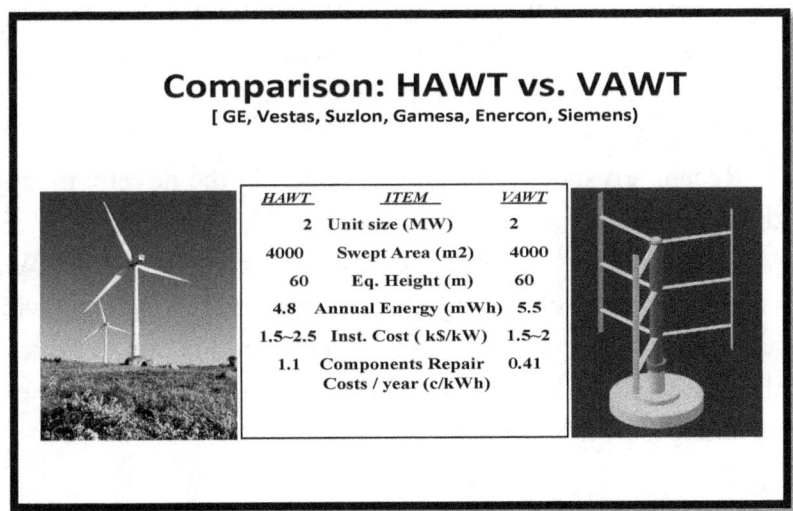

Fig. 4-5 Comparison of HAWT vs. VAWT

The comparison is based on both HAWT and VAWT units having the same design criteria such as:

- ➤ Unit size in MW rating.
- ➤ Equatorial hub height.
- ➤ Swept area.

The results indicate that a straight bladed H-type VAWT unit would be competitive in:

- ➤ Annual energy production.
- ➤ Total installed cost.
- ➤ Annual maintenance cost.

Several cost and technical advancements have been made, based on design improvements introduced, such as:

- ➤ Blade design.
- ➤ Shape and configuration of the unit.
- ➤ Cantilever design (without guy cables).
- ➤ Optimization of system components.
- ➤ Advances in technical analysis.

Recent advances being incorporated in the development of the third-generation of the large MW H-Type, Straight Bladed VAWT with shrouds show promise of keeping the VAWT cost competitive. Such design improvements, together with the advances made during the 1980s improved the competitiveness of the VAWT in the market place. There are inherent existing advantages of the VAWT design.

It will also help to maintain the comparative lead position it had achieved in the 1980s. The overall simplicity of "Design & Construction" of the VAWT consists of:

> ➢ A Vertical-Axis wind turbine with multiple rotors and a supporting frame is simple in design and construction.
> ➢ An H-type, straight- bladed Vertical-Axis wind turbine with multiple rotors, is rugged in construction.
> ➢ A Power train design located at ground level provides reliability and safety in operation, and easy access to all power train equipment.

As illustrated in the Figure 4-6, additional improvements in the near future such as a shrouded design would further enhance the ability of the VAWT to be competitive with other technologies for the generation of electricity.

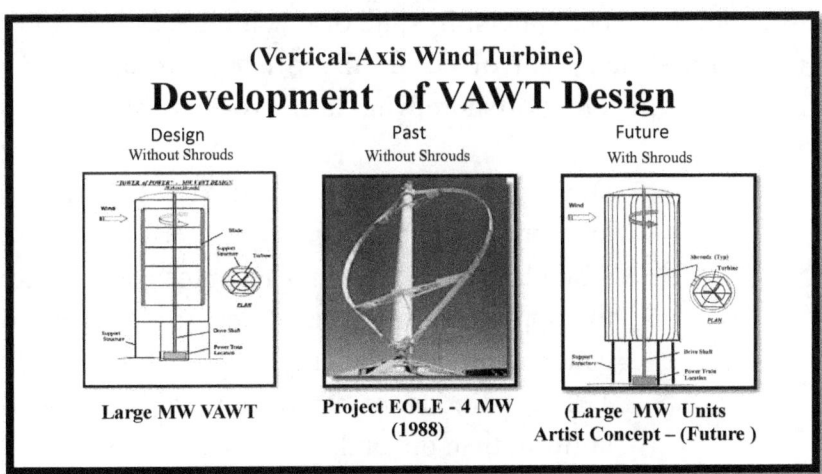

Fig. 4-6 Development of VAWT design

Recent advances being incorporated in the development of the third-generation of the large MW H-Type, Straight Bladed VAWT with shrouds show promise of keeping the VAWT cost competitive.

4.10   OFFSHORE MW WIND TURBINES

The following provides an overview of offshore installations.

The first offshore HA wind turbines were installed off the coast of Denmark in 1991.

There are several countries in Europe that now have offshore wind farms operating in shallow waters. Countries, such as UK, Denmark, Norway, Belgium and Germany have installed large wind turbine farms off their shorelines.

Research work on advanced turbine design and foundation technologies are being developed to install wind farms further offshore in deeper waters. In spite of much higher costs, (about 3 to 4 times more) involved in the offshore installations, the reasons for pursuing offshore wind farm projects are:

> - Does overcome the "NIMBY" syndrome, thus avoiding the public concerns.
> - Abundant offshore wind resources with potential to supply enormous quantities of electrical power generated by wind energy.
> - The conditions of wind flow at sea are more uniform than on land.
> - Higher wind speeds at sea mean higher annual energy production making it cost effective.

Statistics on offshore wind farm installations indicate that by the end of 2012, about 1,662 turbines at 55 offshore wind farms across 10 European countries were generating electricity enough to power almost five million households.

By January 2014, 69 offshore wind farms had been constructed in Europe with an average annual rated capacity of nearly 500 MW in 2013 and the total installed capacity of offshore wind farms in European waters had reached over 6,500 MW. Projections for 2020 calculate a wind farm capacity of 40 GW in European waters, which would provide about 4 ~ 5 % of the European Union's demand for electricity.

The Chinese government has set an ambitious target of 30 GW of installed offshore wind capacity by 2020.

India is looking at the potential of offshore wind power plants, with a 100 MW demonstration plant being planned off the coast of Gujarat.

The Norwegian wind industry is predicting power production of about 10 TWh per year. It is a significant milestone considering that Norway set a wind power production record of 2.5 TWh.

In the USA, five wind turbines installed in the summer of 2016, at the "Deepwater Wind Farm" project off Block Island, Rhode Island, mark the first "steel in the water" for U.S. offshore wind power.

There are several countries currently engaged in the development and commercialization of offshore wind farms, and many are considering utilizing a floating foundation technology.

There is a growing interest for floating offshore wind farms at deeper water sites in the USA. There are several projects on-going in France, Japan, Portugal Scotland, and Spain to install wind farms utilizing floating foundation technology.

When offshore wind farms utilizing floating foundation technology are commercialized, and reductions in the costs for projects achieved, then offshore wind farms utilizing floating foundations will gain acceptance in a number of markets for offshore wind farms.

The Figure 4-7 shows a typical offshore HAWT project in operation.

Fig. 4-7 A typical offshore wind farm with HAWT units

4.11   SUMMARY

✓   The developments in fossil fuel systems in the late 1940s virtually eradicated the development work on any large wind turbine systems.

✓   It was only after the energy crisis in the early 1970s, when oil prices increased from 2 US\$ /barrel to nearly 18 US\$/barrel, that the interest in the development of large wind turbines was renewed worldwide.

✓   The two main categories of wind turbines are: Horizontal-Axis Wind Turbine (HAWT), and Vertical-Axis Wind Turbine (VAWT). The technologies for both the HAWT and the VAWT have matured, and are now proven technologies.

✓   However, the current wind power technology of both Horizontal-Axis (HA) and Vertical-Axis (VA) MW units, have inherent design limitations.

✓   Therefore, they are experiencing resistance from the public, and there is a growing rise of "Not In My Back Yard" (NIMBY) syndrome.

✓   In order to overcome the "NIMBY" syndrome, the wind turbine industry needs to overcome the inherent design limitations of wind turbines.

✓   A paradigm shift in its approach to designing wind turbines is essential to make wind turbines "acceptable", "reliable" and "affordable".

CHAPTER 5

SMALL
WIND TURBINE
TECHNOLOGY

5.1   OVERVIEW

Wind turbines, with an electrical power rating of 100 kW and less (<100 kW) are classified as Small Wind Turbines (SWT).

Depending on the position of the turbine rotor shaft with respect to the ground, either horizontal or vertical, the turbines are categorized as Horizontal-Axis Wind Turbine (HAWT) or Vertical-Axis Wind Turbine (VAWT).

Small Wind Turbines are designed to power homes, farms, and small businesses. With these simple, yet high-tech small wind turbines, individuals can generate their own power and cut their energy bills while helping to protect the environment.

A typical residential wind energy system might be 5 ~10 kW in capacity mounted on an 18 to 24 meter (60-80 feet) support tower. Such a system can meet the electricity needs of a household, farm, or small business.

Small wind turbines may be set up as stand-alone systems, to complement a solar photovoltaic (PV) system, or be interconnected with the utility grid.

There are many regions where wind conditions are favorable and where fuel costs and maintenance costs for power generation with conventional and non-renewable fuel resources are high (in the range of 25 to 100 c/kWh and higher), due to the cost of fuel transportation to the site (depending on the site location).

In such areas, small wind turbines with ratings in the range of 5 kW to 100 kW offer a cost-effective alternative to conventional methods of power generation.

5.2   DESIGN OF SMALL HAWT

The following Figure 5-1 illustrates the design of a typical Small Horizontal-Axis Wind Turbine (HAWT), with a direct-drive generator.

Fig. 5-1 Typical Small Horizontal-Axis Wind Turbine

There are seven main parts to a horizontal-axis wind turbine:

1. The Base.   The base, generally made of concrete and steel, supports the wind turbine structure. The tower is fitted with a base flange, which is normally attached to the foundation by anchor rods embedded into concrete and bolted to the implanted tower stub. For the foundation, depending on the condition of the ground at the site, where the turbine is being installed, various designs of concrete slabs, with multi-pile and mono-pile solutions have been used for the support tower.

2. Support Tower The turbine is mounted on top of a tall tower to allow the blades to take advantage of the higher wind speeds at an elevated level. For small wind turbines, the support tower elevates the turbine rotor to the desired height in the range of 18 to 40 meters. There are three types of support tower, namely: (1) Guyed towers, lattice/monopole; (2) lattice towers, guyed or self-supporting; and (3) monopole towers, guyed or self-supporting.

3. Nacelle The nacelle houses a generator and gearbox, and other essential drive train system components. The gears increase the low rotational speed of the blades to the generator speed of 1,500 to 1,800 RPM. Rotation of the generator produces electricity, which is fed onto the electrical grid system. Certain wind turbines use a direct drive system. A direct drive system connects the rotor directly to a generator.

4. Blades Blades have an aerodynamic profile, like an airplane wing. Wind turbine blades use lift to capture the wind's energy, by spinning a generator in the nacelle. The blades of a small wind turbine, depending on the size of the turbine, spin at a rate of about 150 to 300 revolutions per minute (RPM), with the speed at the blade tip as high as 150 to 200 miles per hour. Depending on the size of the unit, the length of blades for a small wind turbine varies from less than one (1) meter for a unit of 1 kW or less to fifteen (15) meters in length for a unit of 100 kW electrical power rating. Normally, most of the modern day blades are made of materials that have high strength-to-weight ratios, such as fiberglass.

5. Rotor   Generally speaking, the rotor for a typical small wind turbine consists of three blades, a hub, and a low speed shaft. The blades with an aerodynamic profile generate lift, which makes the turbine rotor turn. They are bolted onto the hub, with a pitch control mechanism to permit the blade to rotate about its axis, and help to limit the rotational speed of the rotor under varying wind speeds.

6. Drive-Train System   The wind turbine rotor drives a shaft, which connects to a gearbox, which increases the revolutions per minute to a speed suitable for the electrical generator. Electrical power generating is the main system of the wind turbine. The gearbox should be strong enough to handle the frequent changes in torque caused by variations in the wind speed. Wind turbines have either induction or permanent-magnet generators, depending on the type of wind turbine drive-train system design.

7. Yaw System   All horizontal-axis wind turbines have a yaw drive system to keep the rotor facing (HAWTs need to constantly align themselves with the wind using a yaw adjustment mechanism) into the wind. They also help to unwind the cables that travel down to the base of the tower. The yaw-drive may consist of a tail boom with tail vane or a system consisting of gears and either an electric or hydraulic motor mounted on the nacelle, which prevents the turbine from turning, and stabilizes it during normal operation.

Installation   Installations at project sites involve transportation of turbine components by road, rail, and water. Once the system components such as the nacelle, tower sections and blades are delivered to the installation site, construction of the wind turbines can start. In addition to the erection of each turbine, there is additional work such as improving access roads, laying electrical cable, installation of an electrical substation, and connecting each turbine to the power grid.

Fig. 5-2 Example of small VAWT unit

5.3   DESIGN OF SMALL VAWTs

Vertical axis wind turbines have been the focus of the Canadian Energy program since its early interest in the technology in 1965. Since then, there have been successful demonstrations of prototype small VAWT units. The VAWT technology has matured.

Advanced second-generation small VAWTs are competitive with commercially available HAWTs. Development and successful operating experiences with VAWTs added many technical innovations and improvements to VAWT technology. VAWT technology is a proven technology, based on its merits of technical superiority and cost competitiveness.

Unlike the horizontal-axis wind turbines, where it is necessary to constantly make adjustments to face the turbine into the wind, as illustrated in the Figure 5-2, the design of the VAWT rotor is very different.

The VAWTs are always aligned with the wind. The turbine does not need to be pointed into the wind to be effective; and there are no adjustments needed when the wind direction changes. Many other components of the VAWT system are similar to the HAWT system.

5.4   BASIC CONCEPT: SINGLE LINE DIAGRAM

Figure 5-3 illustrates a basic concept of small hybrid wind solar power system and its various applications. Except for a stand-alone system, in all other applications, the requirement for having an energy storage system is not essential.

Because of the intermittent nature of wind energy, energy storage becomes essential only if it is required to be a source of "On-Demand" power supply. The amount of energy storage needed for the application depends on:

> Availability of wind and solar resources at the installation site.
> Duration of on-demand power supply requirements during periods of unavailability of energy.

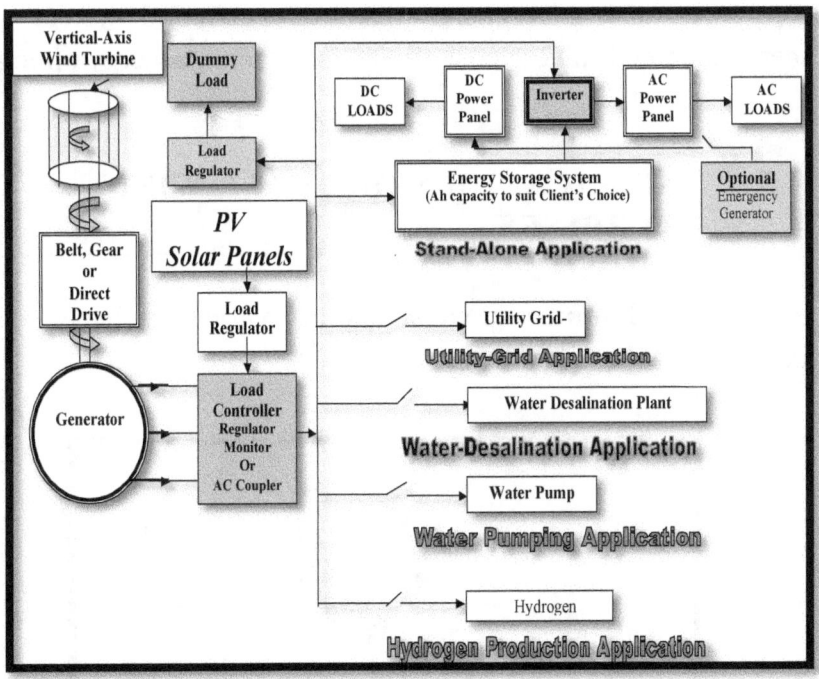

Fig. 5-3 Single line diagram

In a stand-alone application, to guarantee the requirements for an on-demand power supply, the costs of the energy storage system utilizing batteries can be substantially reduced by the installation of an emergency generator using fuel, either diesel or gasoline.

At present, due to the rapid progress being achieved in the development of battery systems, all indications are that energy storage systems utilizing batteries are poised to change the manner in which we select and manage our choices in an on-demand power supply. Recent advancements in battery technology, coupled with substantial decreases in the price for batteries, continue to improve the viability of installations of hybrid wind and solar systems. In many locations blessed with good wind and solar energy resources, a hybrid system with energy storage is a very cost effective and viable energy option.

5.5   APPLICATIONS OF SMALL WIND TURBINES

Small wind turbines are available in a variety of shapes, sizes, and electrical power ratings.

Fig. 5-4 Examples of small wind turbines

Small (micro) wind turbines are ideally suitable for electrical power generation in a variety of applications including, but not limited to, applications such as: A hybrid system, consisting of wind, solar (PV), battery storage, invertors, and a back-up diesel generator.

Since then, there have been successful demonstrations of prototype VAWT units, ranging from 30 to 500 kW rating, in many countries including the USA, the UK, Netherlands, Sweden, Germany, and China, and with units in the range of 50 to 4,000 kW in Canada.

A hybrid system, consisting of wind, solar (PV), battery storage, invertors, and a back-up diesel generator, is ideally suited to offer the best solution to suit the energy profile and the specific need of most applications. Wind/Solar hybrid systems benefit from each other's technology assets.

To ensure an "uninterruptable round-the-clock" electrical power supply, a small hybrid Wind/Solar system generally consists of a wind turbine for generation of electricity, a battery bank for power storage, with PV solar panels, and a diesel or propane generator for emergency power supply.

In many locations, wind and solar resources compliment each other, thus enabling designers to reduce the size of each component.

Fig. 5-5 Micro wind/solar hybrid systems

Since the renewal of interest in wind energy in the early 1970s, electrical power generation by utilizing wind turbines of both types, (the horizontal axis and the vertical axis), has matured into proven technologies.

Small wind turbines are available in a variety of shapes, sizes, and electrical power ratings. They can be tailored to suit a variety of applications. Small (< 100 kW) wind turbines are ideally suitable for electrical power generation for a variety of applications such as:

Small (micro: 1 kW and <) for
- ➢ homes
- ➢ cottages
- ➢ telecommunication systems
- ➢ hybrid system
- ➢ street lighting

Small (10 kW and <) for
- ➢ homes
- ➢ farms
- ➢ oil-well operation
- ➢ water pumping
- ➢ rooftops
- ➢ hybrid wind/solar systems
- ➢ village electrification

Small (20 kW and <) for
- ➢ farms
- ➢ oil-well operation
- ➢ water pumping
- ➢ stand-alone power
- ➢ hybrid wind/solar system
- ➢ pipeline protection
- ➢ village electrification

Small (100 kW and <) for
- ➢ hybrid wind/solar systems
- ➢ stand-alone power system
- ➢ farms
- ➢ urban, rural, and remote communities
- ➢ village electrification

5.6   SWT   PROBLEMS AND SOLUTIONS

The main two areas of problems with the existing design of small wind turbines (SWT) are:

> ➢ Reliability of small wind turbines.
> ➢ Acceptability of small wind turbines by the public.

For these two reasons the existing small wind turbine industry is:

> ➢ Unable to meet the existing demand for SWTs.
> ➢ Powerless to capture the growing market for SWTs.

Public concerns with respect to "Acceptability" are specifically related to:

> ➢ Damage or destruction due to lightning strikes.
> ➢ High maintenance costs.
> ➢ High noise emissions.
> ➢ Safety issues.

Damage or Destruction Due to Lightning Strikes   In most of the small wind turbine designs, selection of blade materials in wood or fiberglass was made to make blades lighter in weight and to reduce stress.

However, it did not address the issues related to the blades being subject to damage and destruction due to lightning strikes. Units with blades made of fiberglass, wood or other materials are susceptible to damage by lightning. Warranty on the units is limited. After each case of severe weather condition, inspection of the turbine blades is recommended, (in accordance with operation and maintenance manuals) which results in additional cost for maintenance.

Solution   To prevent the problem of lightning strikes on HAWTs and VAWTs; the units must be designed to provide total lightning protection, thus eliminating the need for periodic shutdown for inspections and maintenance and repair of damage due to lightning strikes. Total lightning protection means less downtime and more energy production.

High Maintenance Costs   Most of the present day HAWT and VAWT units are designed without any provision for a maintenance platform. This results in higher maintenance costs.

Solution   Units that are designed with a maintenance platform for easy access to equipment for inspection and maintenance, result in better reliability and availability of the unit, and therefore reduced maintenance costs and more annual energy production.

High Noise Emissions   Rotational speed, size of the blades, and swept areas of the HAWT designs were optimized in order to reduce production costs and to reduce the cost of energy production. This resulted in high rotational tip-speeds for the turbine blades, leading to the creation of problems with unacceptable noise levels (higher than 40 dB), and thereby limiting the acceptability of the units. High blade tip-speed is the main cause of the noise emitted by the turbines. (Sound level higher than 40 dB)

Solution   To achieve any reduction in noise created by the blades, the rotational blade tip speed must be reduced.

Table 5-1 Examples of noise levels

Sound Levels in decibels (dB)	
0 dB	Near total silence
15 dB 45 dB 40 dB	Whisper Day time Noise Regulation Night-time-Noise Regulation
60 dB	Normal conversation
90 dB	A lawnmower
110 dB	A car horn
120 dB	A rock concert or a jet engine
140 dB	A gunshot or firecracker

Safety Issues   To achieve public acceptance, the wind turbine industry must overcome the safety issues related to small wind turbines. The problems are due to the inherent design limitations of small wind turbines., such as:

✓ Lack of rugged design and construction

✓ Safety in Operation

✓ 100% lightning

Solution   The problems related to public acceptance and safety issues of SWTs can only be resolved by specifically designing wind turbines that address the issues related to "Criteria for Public Acceptance".

5.7 SUMMARY

✓ The worldwide interest in the development of "Small Wind Turbines" was renewed only after the energy crisis in the early 1970s, when oil prices increased from 2 US$/barrel to nearly 18 US$/barrel in the late 1970s.

✓ Small wind turbines, the horizontal axis (HA) and the vertical axis (VA), have now matured into proven technologies.

✓ There are many regions, where wind conditions are favorable, and where fuel costs and maintenance costs for power generation with conventional and non-renewable fuel resources are high.

✓ In such regions, small wind turbines with ratings in the range of 1 kW to 100 kW offer a cost-effective alternative.

✓ Small wind turbines can be gainfully utilized to fulfill the future energy needs of both developed and developing nations. Recent advancements in battery technology, coupled with substantial decreases in the price for batteries, continue to improve the viability of installations of hybrid wind and solar systems.

✓ In many locations blessed with good wind and solar energy resources, a hybrid system with energy storage is a very cost effective and viable energy option.

✓ However, the current wind power technology of both horizontal-axis (HA) and vertical-axis (VA) MW units, have inherent design limitations.

✓ Therefore, small wind turbines are experiencing resistance   from the public. There is a growing rise of "Not In My Back Yard" (NIMBY) syndrome.

✓ In order to overcome the "NIMBY" syndrome, the wind turbine industry needs to overcome the inherent design limitations of small wind turbines.

✓ A paradigm shift in its approach to designing small wind turbines is essential to make wind turbines acceptable, reliable, and affordable.

CHAPTER 6

Market Potential

For

Wind Turbines

6.1   MARKET OVERVIEW

The wind power industry, with its small and large wind turbines, is currently the fastest growing source of electrical power supply. Many environmentalists favour it as a prime alternative to fossil fuels.

The world wind generation capacity more than quadrupled between 2000 and 2006 and it is expected to keep growing at double-digit rates for the next twenty years. Depending on the increase in electricity demand, wind power could supply 10 to 15% of global electricity demand in 2020, and as much as 20 to 30% in 2030.

However, in order to capture its fair share of the market for the world's electrical power-generating capacity mix, the wind power industry, for both large MW and small KW wind turbines, needs to address and overcome the inherent limitations of wind turbines, with regard to "Clients' Concerns".

Market growth for wind turbines is severely curtailed by the growing public concerns with issues related to wind turbines, issues such as: safety, reliability, noise-free in operation, birds and bats killed, and visual aesthetics.

6.2:   MARKET FOR SMALL WIND TURBINES

Utilization of renewable energy resources, such as wind power, is vital for addressing the future energy needs of both developed and developing nations.   There are about 350 manufacturers of small wind turbines currently in operation in over 40 countries globally.

In the year 2008, the resurgence of interest in power generation by small wind turbines was spurred for many reasons. These included:

> Public acceptance ~ use of renewable energy.
> Wind power as an economical, clean, limitless, and sustainable energy alternative.
> Demand for risk-free and acceptable forms of energy.
> Preservation of environment from pollution, & global warming.
> Improvements in life cycle costs for small wind turbines, due to reduction in prices of small wind turbines, escalating costs for conventional fuels, and escalating costs for maintenance of conventional units.

Depending on application, small wind turbines, with ratings in the range of micro (<1 kW) to 100 kW, offer a cost-effective alternative to conventional methods of power generation.

Numerous studies on the global market for small wind systems (those with less than 100 kW of generating capacity) indicate that the market is growing rapidly. Global market potential for small wind turbine systems is expected to maintain double-digit growth and continue a 20% to 30% plus annual growth for the next decade.

The market potential for small wind turbines, by the year 2020, is estimated at about 1 billion US$ / year in the USA, and 2 to 3 billion US$ / year globally. The wind energy industry for small wind turbines is positioning itself as a serious contender for power generation in a variety of applications.

Lately, energy users and planners are looking to wind generated power as an economical, clean, and limitless energy alternative.   At present, appropriate small wind turbine technology is not yet available for lots that are less than one (1) acre. (Due to a lack of public acceptance of the current technology)

The availability of small wind turbines is limited due to the inadequate design of the machines. Clients are demanding machines with technical features that offer added value to the customer, features such as: reliability, ruggedness, addressing of social concerns about safety and noise, longer life expectancy, lightning protection, and access for ease of inspection and maintenance. However, the problems with all existing designs of small wind turbines are: acceptability, reliability, and affordability. For these reasons, the existing small wind turbine industry is:

> Unable to meet the existing demand.
> Powerless to capture the growing market, for small wind turbines.

The growth of market potential will be fuelled by the impact of population explosion. It will accelerate and expand the demand for small (<100 kW) size wind turbines. Fluctuations and impact of the rise in crude oil prices is much more dramatic. It makes the economical application of small wind turbine installations imminent and cost competitive with electrical power generation utilizing oil as fuel.

The total global market potential for small wind turbines by the year 2020 is over 210,000 MW. The market potential for small wind power systems is expected to maintain double-digit growth rate and continue a 20% to 30% plus annual growth for the next decade.

Nowadays, there is a global consensus and political will to get off the addiction to oil and adopt the utilization of renewable energy resources. In addition, it provides sustainable energy with environmental benefits by reducing pollution.

Worldwide, over 2 billion people in the developing world do not have access to electricity. The greatest potential market for small wind turbines is in those parts of rural areas where millions of people do not have access to electricity. Small wind turbines are suitable for electrical power generation for a variety of applications such as:

➤   Stand-alone power systems.
➤   Off-grid applications.
➤   Grid-connected power facilities.

There is a recent trend by many governments to consider offering specific support policies for hybrid wind and solar systems for projects such as village electrification. In such applications where other sources of power would be unreliable and/or more costly, the small wind turbines will be cost-effective.

6.3   MARKET FOR LARGE WIND TURBINES

According to the forecasts for market potential for large multi-megawatt wind turbines, most countries worldwide are planning to ensure that wind energy supplies 20 percent of the country's demand by 2025.

The market potential for wind power systems is expected to maintain double-digit growth and continue a 30% plus annual growth for the next decade. In 2015 total investments in the clean energy sector reached a record of about USD 330 billion.

In early 2015, expectations for growth in the wind power market were not excessive, because of continued economic slowdown in Europe and some emerging markets.

The political uncertainty in the US made it difficult to make projections for 2015. In addition, this did not factor in the ability of China to surpass all projections with exceptional wind power development numbers. By the end of 2016, eight countries had more than 10 GW of installed capacity including:

- China (145 GW)
- U.S.A. (74 GW)
- Germany (44 GW)
- India (25 GW)
- Spain (23 GW)
- UK (13 GW)
- Canada (11 GW)
- France (10 GW)

China crossed the 10 GW mark in 2014, adding another chapter to its already exceptional history of renewable energy development since 2005. This year it made history again and strengthened its position on the leaderboard.

It should be noted that these double-digit growths in installations of wind turbines were achieved in spite of the problems of inherent limitations associated with the designs of wind turbines. Installed capacity of wind turbines that are designed to address the public's concerns will certainly improve the market growth for wind turbines.

The limitations that concern the clients and the public most, which lead to "NIMBY" syndrome, are safety in Operation, Lightning Strike, Noise, and birds/bats killed.

The global cumulative installed capacity of wind turbines, and the predictions of the potential market for wind turbines are illustrated in Figure 6-1.

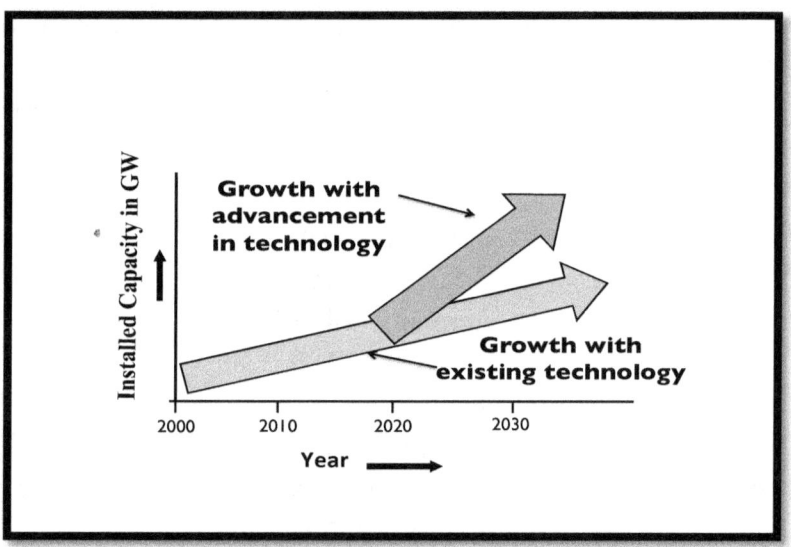

Fig. 6-1   Potential Market for wind turbine industry

> With further advancements in technology and future improvement in the designs of wind turbines, the potential for market growth rate is at least five times the current level.

> Annual revenues for the service-industry for wind farms are anticipated to increase to about $15 billion by the year 2020.

➢ Worldwide, there are about 700,000 to nearly 1,000,000 people who are employed in the wind industry growth in wind power is expected to grow in double digits.

➢ Most of the installations of wind farms are onshore. Globally, offshore wind farm installations are being seriously considered.

Renewable energy resources delivered about 17 percent of electricity generation in the U.S. during the first half of 2016. The small wind turbine market in the U.S. grew about 50% in the year 2010.

China   China set a target of expanding wind power capacity to 100,000 MW by the year 2020. China approved the Renewable Energy Law, which mandates power grid companies to buy the entire output of registered renewable energy producers in their areas.

India   India stands at the 5th Rank in terms of total installed wind power capacity just behind China, USA, Germany and Spain. India has plans to achieve an overall renewable energy installed capacity of around 70,000+ MW by 2022.

Current Global Investment:

➢ Large MW units = 100 billion + $ / year.
➢ Small Wind Turbines = 2 ~ 3 billion $ / year.

Future Global Investment:

➢ Large MW units = 300 billion + $ / year.
➢ Small Wind Turbines = 20 ~ 30 billion $ / year.

6.4 SUMMARY

- ✓ Lately, energy users and planners are looking to wind generated power as an economical, clean, and limitless energy alternative.

- ✓ **Small Wind Turbines:** The wind energy industry for small wind turbines is positioning itself as a serious contender for power generation in a variety of applications. The total global market potential for small wind turbines by the year 2020 is over 210,000 MW. Global market potential for small wind turbines, by the year 2020, is estimated to be ~ 2 to 3 billion $ / year.

- ✓ **Large MW Wind Turbines:** The global wind energy market was worth $130 billion in 2013 and about $165 billion in 2014. The market is expected to grow at a compound annual growth rate resulting in about $250 billion by the year 2020.

- ✓ Depending on the country, attaining 20% of the country's electricity supply from wind by 2025 would generate around 64 to 600 billion US $ of investment, However, in order to capture its fair share of the market for the world's electrical power-generating capacity mix by large MW wind turbines, the wind power industry needs to address and overcome the inherent design limitations of wind turbines.

- ✓ Wind turbine design must address "Clients' Concerns" such as: safety, reliability, noise-free in operation, birds and bats killed, and visual aesthetics. Wind turbine systems must be designed to be "acceptable", "reliable", and "affordable".

CHAPTER 7

GLOBAL WARMING

AND

ENVIRONMENTAL

ISSUES

7.1   OVERVIEW

In an energy-hungry world, to meet our current demand and our future energy needs, all nations must utilize all available energy resources, including oil, coal, natural gas, nuclear, hydro, wind and solar.

However, due consideration must be given to global warming and climate change; issues that threaten the quality of life for the present and future generations to come.

If we are to avert the catastrophic consequences of global warming, then it is imperative that we demand the shift from conventional fossil fuel to clean, and sustainable renewable energy sources.

As the public becomes more conscious that sustainability and energy savings can be achieved by utilization of renewable energy resources, then generation of electricity by renewable energy resources will be of paramount importance.

At present, the development of the electricity supply sector is characterized by a dynamically growing renewable energy market and an increasing share of renewable electricity.

This will compensate for the phasing out of power plants utilizing conventional non-renewable energy resources.

If urgent actions are taken now, then by 2050, about 70% of the electricity produced worldwide could come from renewable energy sources such as wind, solar thermal power, photovoltaic, and other renewable energy resources.

7.2   GLOBAL WARMING AND ITS IMPACT

Greenhouse gases like methane and carbon dioxide cause global warming. Global warming is the result of over-production of these gases. The impact of an increase in the amount of greenhouse gases results in increasing the earth's average surface temperature.

Persistent build-up of greenhouse gases in the earth's atmosphere is causing serious disruption in the ecosystems. The main greenhouse gas is carbon dioxide (CO2) produced by using conventional non-renewable fossil fuels for energy and transport.

Fig. 7-1 Pollution from coal-fired power plants

Global warming in recent decades and its projected continuation is the observed increase in the average temperature of the Earth's atmosphere and oceans. If rising temperatures are to be kept within acceptable limits then we need to significantly reduce our greenhouse gas emissions.

This makes both environmental and economic sense. In order to understand global warming and environmental issues with electrical power generation, it is essential to have a clear understanding of the following:

> Impact of population growth, on global warming.

> The availability of conventional non-renewable
  and renewable energy resources, and options to
  meet the future energy needs for electricity.

> Impact on "social costs" and "environmental externalities" to give a perspective on the cost of energy resources.

The need for "political will" to implement immediate actions to avert the foreseeable disasters from global warming are obvious.

If the increase in global temperatures continues, it can in turn cause changes, including a rising sea level and changes in the amount and pattern of precipitation.

If rising temperatures are to be kept within acceptable limits then we need to significantly reduce our greenhouse gas emissions. This makes both environmental and economic sense.

These changes may increase the frequency, and intensity of extreme weather, and its consequences: heat waves, droughts, disease, shortages of food, and fresh water.

An explosive rise in world population from 2.5 billion in 1950 to over 7 billion in the year 2015, combined with adopting changes in life styles has resulted in causing severe air pollution and global warming.

Figure 7-2 shows the problems related to climate change that could result in global warming with serious consequences.

Fig. 7-2 Problems related to climate change

Combined with adopting changes in life styles based on industrialization and consuming increasingly larger amounts of energy resources has triggered a dramatic depletion and a rise in price of all conventional fuels, such as oil, coal, and nuclear.

Figure 7-3 shows most likely impacts of global warming by the year 2100. If the problems related to climate change are not addressed that could result in global warming with serious consequences.

Impact of Global Warming
by the year 2100

⚙ **2 to 4 billion people without water**

⚙ **200 ~ 500 millions of deaths due to heath**

⚙ **300 ~ 800 millions of deaths due to starvation**

⚙ **Over 100 million/year deaths due to floods**

⚙ **Drastic impact on climate and economy**

Fig. 7-3 Impact of global warming (year 2100)

7.3   REASON WHY WE SHOULD CARE

Scientists and engineers from around the world have compiled evidence that tells us that the planet is warming. With the population increasing to an estimated 9 billion by the year 2050, it is certain to cause an increase in air pollution and global warming.

The reasons for utilization of renewable energy resources for the generation of electrical power supply are, freedom from:

➢ Energy shortages
➢ Global warming
➢ Environmental pollution
➢ Price increases

By utilization of non-renewable energy resources in an indiscriminate manner, we the people have created the problem, and now that we understand the problem, it is up to us to resolve it.

Utilization of renewable energy resources for electrical power supply will help us achieve:

➢ Freedom from conflicts - security of supply
➢ Freedom from global warming
➢ Freedom from pollution
➢ Freedom from energy shortages
➢ Freedom from price increases

What is not a solution?

➢ More taxes
➢ Smaller cars
➢ More research, more studies
➢ Long-term solutions

What is needed to resolve the problem is:

➢ Changes in life styles
➢ Practice of energy conservation
➢ Utilization of sustainable renewable energy resources

> ➢ Leveling of the playing field for all renewable and non-renewable energy resources

If we are to avert the catastrophic consequences of global warming, then it is imperative that we demand the shift from conventional fuels to renewable energy sources.

Figure 7-4 suggests solutions to combat global warming.

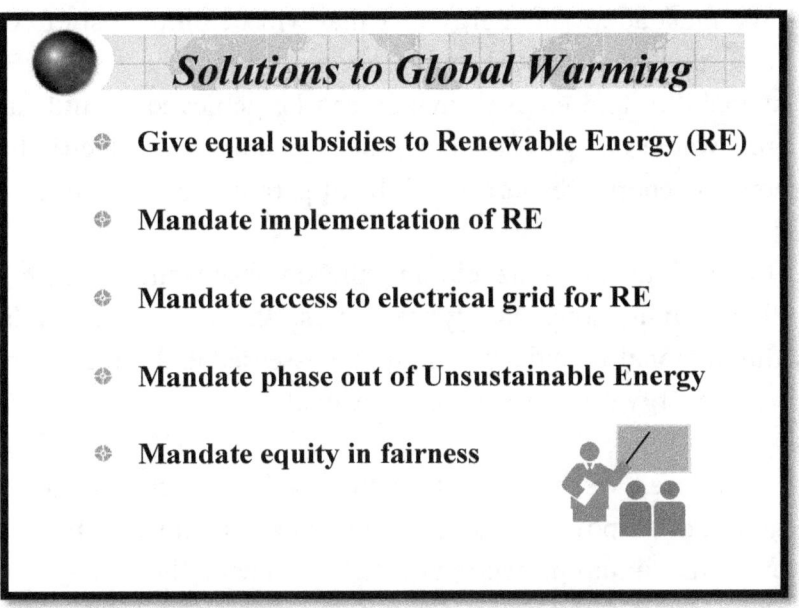

Fig. 7-4 Solutions to global warming

Renewable resource utilization should focus on a future-looking systems approach to deliver abundant, clean energy to all sectors of the economy. This would yield maximum national impact by reducing imports of resources, improving balance of payments, preserving the environment from pollution and providing better protection of consumers.

In the next few decades, the availability of energy and the cost of energy are likely to remain the two most widespread issues facing both developing and developed nations.

Improvements in energy supplies in developing nations will require promotion of the use of indigenous renewable resource systems as an alternative to systems utilizing conventional non-renewable energy resources.

We must adopt an energy mix that will be more cost effective and sustainable. As the public becomes more conscious that sustainability and energy savings can be achieved by utilization of renewable energy resources, then generation of electricity by renewable energy resources will be of paramount importance

In the interest of developing the domestic energy base and utilizing indigenous energy resources, the resources available within national boundaries should be assessed and the economics of such energy options should be evaluated.

Therefore, we must utilize the available technologies for generation of power in a safe and clean manner not only to provide maximum protection of the consumers, the taxpayers and the ratepayers, but also to ensure that we do not pass the burden of cleanup and waste disposal as our legacy to future generations.

7.4   ENERGY SUPPLY MARKET FORCES

The three major characteristics for energy supply market forces are:
- Cost of Energy
- Social Awareness
- Governmental Policies

Cost of Energy: Whereas the prices for conventional fuels, such as oil, gas, coal, and nuclear fuels are getting extremely volatile and rising, the price for renewable energy is steadily going downward.

Social Awareness: In response to public awareness of global warming and issues related to climate change, many countries are adopting government-sponsored programs and setting up regulations and standards for emission control so that environmental friendly energy resources can be utilized.

Governmental Policies: Energy security, environmental protection and availability of clean and sustainable energy resources are the key political concerns, which favor the use of renewable energy resources.

With large-scale production and competitive costs, commercial application of renewable energy such as wind and photovoltaic is possible. Figure 7-5 compares environmental impacts of electrical power generation by various energy resources.

As illustrated in Figure 7-5, each such system has advantages and disadvantages but many of them pose environmental concerns. Because of the large amounts of electricity consumed by present day society, the influence of electricity generation on the environment is significant.

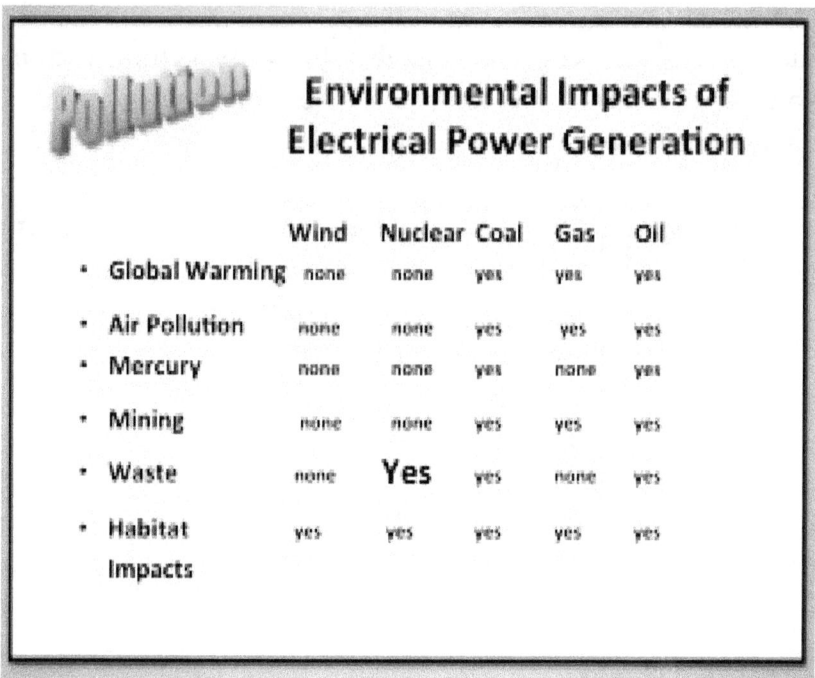

Fig. 7-5 Impacts of electrical power generation

. If urgent actions are taken, the goal of about 70% of the electricity produced worldwide by 2050 from renewable energy resources is achievable

7.5   ENVIRONMENTAL BENEFITS

Every unit (kWh) of electricity produced by utilizing renewable energy resources, such as wind and solar, replaces a unit of electricity that would otherwise have been generated by a power station burning fossil fuel.

Estimates of total global warming emissions of wind turbines depend on factors such as: (a) construction materials of the wind turbine, (b) wind speed, and (c) the percent of time the wind is blowing.

Life-cycle global warming emissions of wind turbines are estimated to be between 0.02 and 0.04 pounds of carbon dioxide equivalent per kilowatt-hour.

In comparison, estimates of "life cycle global warming emissions" for coal-generated electricity are between 1.4 and 3.6 pounds of carbon dioxide equivalent per kilowatt-hour, and estimates for natural gas generated electricity are between 0.6 and 2 pounds of carbon dioxide equivalent per kilowatt-hour.

A metric measure used to understand the carbon benefits of wind energy is that each 1.5 MW wind turbine removes about 2,500 ~ 3,000 metric tons of CO2 per year, or the equal of planting about 3 ~ 5 square kilometers of forest every year. (Depending on wind regime at the site)

Every 1,000 MW of wind power installed can reduce CO2 emissions by approximately 3 million metric tons (MMT) annually if it replaces coal-fired generating capacity, or about 2 MMT if it replaces generation from an average fuel mix. For every unit (kWh) of electricity produced from coal-fired plant, the typical emissions are:

- ➢ 750 ~ 1250 grams CO_2/kWh
- ➢ 5 ~ 10 grams SO_2/kWh
- ➢ 3 ~ 6 grams NO_X/kWh

Therefore, each year, every kWh produced by utilizing renewable energy resources such as wind and solar power will contribute to emission reductions. Emissions avoided using wind energy significantly reduces polluting substances emitted by conventional electric power stations.

Table 7-1: Pollution Prevention by Small Wind Turbines

Emissions	Pollution Prevented by Small Wind Turbines (In metric tons/year)			
	Wind Turbine System			
Power Rating (KW)	1	10	100	100
Carbon Dioxide (CO_2)	1.5 ~3.3	12 ~ 32	85 ~220	850 ~220
Sulphur Dioxide (SO_2)	0.01 ~0.025	0.15 ~0.30	0.65 ~1.65	6.5 ~ 16.5
Nitrous Oxide (NOx)	0.07 ~0.015	0.05 ~0.09	0.4 ~ 1.0	4 ~10

As indicated in Table 7-1, each year, a typical wind turbine rated capacity of 1 kW to 1000 kW will contribute emission reductions. (Depending on wind regime at the site.)

Comparable to coal, nuclear power also produces very serious environmental impacts, although indirectly. Whereas nuclear power plants do not release toxic chemicals similar to conventional power plants using coal, oil, or natural gas; nuclear fuel systems do pose risks of catastrophic accident.

It does generate hazards that may well threaten people and damage the environment now and for centuries, and for generations to come. Mining, processing and transporting nuclear fuel produce a vast amount of pollution, including air pollution.

After multiple decades of nuclear power plant operation worldwide, no country has been able to solve the problem of how to get rid of the nuclear waste, or found any solution of safely storing hazardous nuclear wastes for centuries to come, or addressed the questions of how much it will cost and who will pay for it.

7.6   SUMMARY

✓ By the rapid advancement and development of renewable energy technologies, it is economically feasible to cut global CO_2 emissions by almost 50% to 70% by the year 2050.

✓ In the meantime, all nations must utilize all available energy resources, including oil, coal, natural gas, nuclear, hydro, wind and solar.

✓ However, due consideration must be given to global warming and climate change; issues that threaten the quality of life for the present, and future generations to come.

✓ We must ensure that we do not pass the burden of cleanup and waste disposal as our legacy to future generations.

✓ Use indigenous renewable resource systems as an alternative to systems utilizing conventional non-renewable energy resources.

✓ When a fair evaluation is made in comparing the available options for power generation, then, wind and solar power can compete with production of electricity by the utility grid systems with conventional non-renewable fuel energy resources.

CHAPTER 8

COSTS AND ECONOMICS OF SMALL WIND TURBINES (< 100 KW)

8.1   OVERVIEW

Since 1974, the development of the small wind turbine industry has progressed more than in the preceding 100 years. At present, the small wind turbine industry is sufficiently matured to mass-produce its small wind turbine products. By 1984, the concerns for implementation of wind power were not technical, but related to their cost and economic competitiveness with conventional power plants using non-renewable energy resources.

Because of the intermittent nature of the energy resource, a combination system of wind and solar (hybrid) power with energy storage offers the best solution for applications such as stand-alone power systems and village electrification. As shown in the following Figure 8-1 the cost of energy from renewable and especially wind and solar power is rapidly improving.

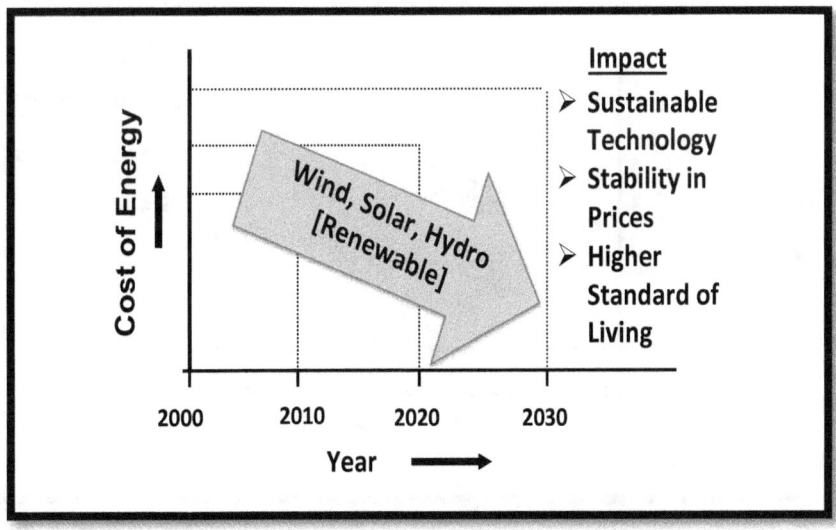

Fig. 8-1 Cost of renewable energy by the year 2030

In many areas with adequate wind conditions and lack of availability of a utility grid system, small wind turbine systems and especially wind/solar hybrid systems are technically and economically viable as an alternative to conventional non-renewable fuel powered generating units.

Due to advancements in technology and economy of scale based on mass production, substantial improvements have been achieved in lowering the cost of small wind turbines and solar (PV) panels. Costs vary depending on the country of origin. For example:

➢ Cost of wind turbine in the late 1970s = 5~8 $/W, and in the year 2016 = 1~2 $/W.

➢ Cost of solar panels in the late 1970s = 6~7 $/W, and in the year 2016 = 0.75~1 $/W.

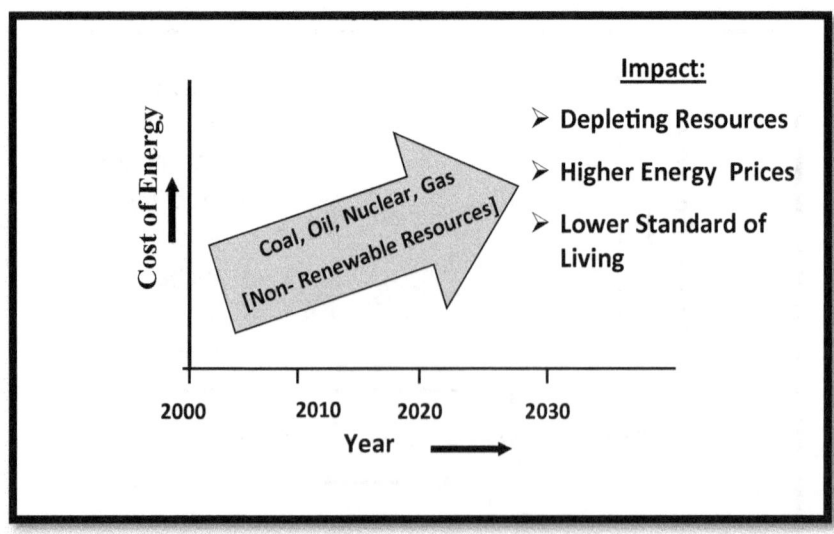

Fig. 8-2 Cost of non-renewable energy by the year 2030

Future advancements in energy storage systems will further enhance the economics and boost competitiveness of small wind turbines with power generating units using non-renewable energy resources. Since the energy crisis in the early 1970s, as shown in the Figure 8-2, the cost of energy of non-renewable energy resources has been on a continual rise.

In addition to fuel costs, there has been a substantial increase in the cost of power plants for generating electricity utilizing conventional non-renewable energy resources including coal and nuclear. Costs vary depending on the country of origin. For example:

➤ Cost of coal plants in the late 1970s = 0.6~0.8 $/W, and in the year 2016 = 4~8 $/W.

➤ Cost of nuclear plants in the late 1970s = 1.5~2.5 $/W, and in the year 2016 = 5~8 $/W.

8.2   SMALL WIND TURBINE COSTS

Costs of Small Wind Turbines (SWT) depend greatly on economies of scale. Costs also vary depending on the country of origin.

Often times there are tax and other incentives that can dramatically reduce the cost of a wind project. The following costs are based on pricing estimates in units made in North America.

The cost may vary based on the country of origin (manufacturing), the location of the installation and other site-specific factors.

Table 8-1 provides an overview of typical costs of Small Wind Turbines (SWT of < 100 kW).

Table 8-1 Costs of small wind turbines

Item	Typical Application		
	Battery Charging	**Housing**	**Industrial Commercial Farms**
Power Rating (kW)	1 kW & <	10 kW & <	100 kW & <
Capital Cost ($)/kW	2,000 ~3,000	3,000 ~ 4,000	2,000 ~ 2,500
Total Installed Cost: ($)/kW	3,000 ~ 4,500	4,500 ~ 6,000	3,000 ~ 4000
O & M Cost: ($)/year	40 ~130	1,000 ~ 2,000	3,000 ~3,500
Cost Of Energy (c/kWh)	Varies with location and Wind conditions at the site		

8.3 ECONOMICS OF SMALL WIND TURBINES

The "Cost Of Energy" (COE) of the electrical power produced by the wind turbine is calculated by the formula:

COE = [(CC * FCR) /AE] *100 + (O&M)/AE *100 - in cents/kWh, where:

CC =   Capitol Cost (Total installed Cost) - in US$

FCR =   Fixed Charge Rate – in %

AE =   Annual Energy production – kWh/year

O&M =Annual Operation & Maintenance Cost – in US$

CC   Capitol cost is the total amount of money invested in installing the system. It is the investment that is written off over the entire useful life span of the system installation.

O&M   The annual operation and maintenance costs (O&M) are the annual average of the total operation and maintenance costs for the life of the project, scheduled and unscheduled costs for the project including inspections, parts and materials, and labor costs. With improvements in design, the maintenance costs have been drastically reduced to from 2% and 3% of the capitol cost in the 1980s to less then 1% in 2014.

AE   The annual energy production for the wind turbine system represents the average annual kilowatts delivered during the lifetime of the system installation. It takes into account the overall availability and capacity factor of the wind turbine system.

FCR   It is expressed as a fraction of the total installed cost, and it is called the "Fixed Charge Rate (FCR). The fixed charge is the capital-annualizing factor, which is a function of financial parameters, and accounts for the design life of the plant, depreciation, taxes, return to investors, insurance and inflation.

Table 8-2 fixed charge rates

	Fixed Charge Rate		
Life (years)	Interest Rate [%]		
	5 %	7.5 %	10 %
10	0.130	0.146	0.163
15	0.0963	0.113	0.132
20	0.0802	0.981	0.118
25	0.0710	0.0897	0.110
30	0.0650	0.0847	0.106

For example, the "Cost of Energy" (COE) of a 10 KW wind turbine with a capitol cost worth $40,000, 24,480 kWh of annual energy production in a 6 m/s annual mean wind speed, and 30 years of service life, requiring maintenance costs each year of 2% of total installed cost ($ 800), with money at 5% interest rate, would be:

COE =   [(CC * FCR) /AE]   *100 + (O&M)/AE *100 in cents/kWh

COE =   [(40,000 x 0.065) / 24,480] x 100 + (800) / 24,480 x 100

COE =   10.62 + 3.27 = 13.89 cents/kWh

From an owner's perspective, depending on location and site conditions, the installed system with a small wind turbine would ensure that the electrical power is procured at the lowest possible cost per unit of energy. At present, in areas with good wind conditions, small wind turbine systems are cost competitive.

The following Table 8-3 indicates an overview of average cost of electricity for household consumers in various states in the USA, and the general cost of electricity in several countries in Europe.

Table 8-3 Costs of electricity (COE)

Location	Electricity ~ US c/kWh (Year 2015 ~ 2017)
Cost of Electricity (in USA)	
New York	15 ~ 18
California	17 ~ 20
Alaska	20 ~ 22
Hawaii	28 ~ 32
Cost of Electricity (in Europe)	
Denmark	33 ~ 36
Germany	33 ~ 36
Italy	24 ~ 28
Spain	24 ~ 27

Comparatively speaking, COE from an electrical utility grid system varies, depending on the period of electricity demand (Off-Peak & On-Peak), and on location, country, province, and/or state.

In the U.S.A, COE to the consumer varies from 8 to 30 US c/kWh in Off-Peak and On-Peak hours. In Canada, COE to the consumer varies from 8 to 12 c/kWh in Off-Peak to 15 to 20 c/kWh during On-Peak hours.

In remote communities in Canada, COE from diesel power plants in the local grid varies, depending on location, from 40 to 120 c/kWh.

The solution to the problems related to the comparative "Costs and Economics of Small Wind Turbines" lies in evaluating the comparative cost of energy (c/kWh) impartially, by taking into account all costs, involved in the generation of electricity, both direct and indirect including all "Hidden" costs, such as:

➢ Environmental impacts.
➢ Government subsidies, direct and indirect.
➢ Waste disposal.
➢ Decommissioning.
➢ Governmental protection against cost overruns.
➢ Insurance & risk of accidents.
➢ Owner's costs.

At present, the "Hidden" costs of electrical power generation by energy systems are not taken into account in evaluating the comparative cost of energy (c/kWh).

8.4 HYBRID WIND AND SOLAR ELECTRIC SYSTEMS

In remote locations, stand-alone systems can be more cost-effective than extending a power line to the electricity grid.

Stand-alone systems are also used by people who live near the grid but are looking for independence from the power provider or to demonstrate a commitment to non-polluting energy sources. Batteries, and other additional equipment, are required with stand-alone home energy systems. The cost of extending a power line to the electricity grid varies from $10,000 to $30,000 per kilometer.

Successful stand-alone systems generally take advantage of a combination of techniques and technologies to generate reliable power, reduce costs, and minimize inconvenience.

Many hybrid systems are stand-alone systems. They provide power through batteries and/or, when needed, by an engine generator powered by diesel fuel. Depending on the system design requirements, energy storage battery banks are normally sized to supply the electrical load from one day to a week.

During low wind or lack of availability of solar power, the diesel generator can deliver power and can also recharge the batteries. The addition of an emergency diesel generator system helps reduce the component size of the hybrid system, and especially the battery banks, which should be adequately sized to supply electrical needs during non-charging periods. Table 8-4 compares the characteristics of wind and solar power systems.

Table 8-4 Characteristics of wind vs. solar power systems

Description	Wind	PV
Energy resource	Wind	Sunshine
Suitable location	Areas with good wind potential	Areas with good daylight potential
Costs & economics	Varies with wind regime at site & location	Varies with daylight at site & location
Environmental issues	Minimal hidden costs	Minimal hidden costs
Availability of power resource for on-demand power supply	Intermittent power resource requires energy storage system after electrical power generation for on-demand power supply	Intermittent power resource requires energy storage system after electrical power generation for on-demand power supply
Market Potential	Residential	Residential
	Commercial	Commercial
	Utility grid	Utility grid
Inherent limitations of existing designs	Public's concerns and demand for "Acceptability"	Costs

Due to advancement in technology, and economy of scale due to mass production, the cost of energy from renewable resources and especially wind and solar power is rapidly improving. Emergence of advancements of an energy storage system, together with drastic reductions in its costs, now make it possible for wind and solar power systems, both small and large MW scale, to be technically and economically viable as an alternative to conventional non-renewable fuel powered generating units.

8.5   BENEFITS OF SMALL WIND TURBINE SYSTEMS

The benefits of the use of electricity in rural areas are far reaching. Availability of electricity in rural areas would improve "quality of life" by providing access to the energy supply required to have better facilities for:

> ➢   Fresh water supply
> ➢   Communication
> ➢   Education
> ➢   Heath care conditions

The costs of providing electricity to small rural communities through grid extension are very high. Therefore it may not be economically viable, because of low potential electricity demand and long distances between the existing electrical grids, due to the location of the rural area. The length of the payback period for the above-mentioned applications depends on:

> ➢   Turbine size and KW rating
> ➢   Quality of wind at the installation site
> ➢   Prevailing electricity rates
> ➢   Available financing and incentives

The key variables in deciding the "cost of grid extension" consists of:

> ➢   Installation of power supply lines, power distribution substation(s), and a low voltage distribution system.

➤ Size of the load to be electrified, distance of the load from an existing transmission line, and the type of terrain to be crossed.

➤ Lack of availability of local technical and management personnel, high transmission losses, and reliability of the power supply in extreme weather conditions.

The small wind turbine would be cost-effective in applications, where other sources of electricity are not available, and power would be unreliable and/or more costly, in applications such as:

➤ Battery charging.
➤ Water pumping.
➤ Telecommunication systems.
➤ Urban and Rural homes.
➤ Farms.
➤ Communities.
➤ Stand-alone power systems.
➤ Village electrification.
➤ Power systems for agriculture and forestry.

Depending on these and other factors, the time it takes to fully recover the cost of a small wind turbine is 5 to 15 years.

8.6   SUMMARY

✓   Cost of energy generated by small wind turbines depends on location and wind conditions at site. Small wind turbines and Photovoltaic (PV) systems are often the least expensive sources of power for remote sites that are not connected to the utility system.

✓   Because of the intermittent nature of the energy resource, a combination of wind and solar (hybrid) systems with energy storage, offers the best solution for applications such as village electrification and a stand-alone power system.

✓   Hybrid systems such as wind/photovoltaic, wind/diesel, and other combinations can often provide the most efficient and cost-effective option for rural electrification, and stand-alone power systems. Diesel generators or batteries can be used as a backup source to meet the demand for power supply.

✓   When a fair evaluation is made in comparing the available options for power generation, then, in regions with good wind resources, wind power can compete with available conventional power generation for the production of electricity.

✓   The real constraints on the selection of the appropriate energy resources for electrical power generation lie, not in the availability of acceptable options but rather, in the political will of utilities and governments to fairly assess the options with a view to providing a risk-free and environmentally acceptable solution.

CHAPTER 9

COSTS AND ECONOMICS OF LARGE WIND TURBINES

9.1   OVERVIEW

Since 1974, the development of wind power has progressed more than in the preceding 100 years. By 1984, the concerns for implementation of wind power were not technical but related to their cost and economic competitiveness with conventional power plants using non-renewable energy resources.

By 2015, the large MW wind turbine industry was sufficiently matured to mass-produce its large wind turbine products. Wind farms with large MW units are now proving to be cost-competitive with all types of conventional power plants utilizing non-renewable energy resources.

The solution to the problems related to the comparative "Costs and Economics of Wind Turbines" lies in evaluating the comparative cost of energy (c/kWh) impartially, by taking into account all costs, both direct and indirect including miscellaneous (hidden) costs such as: environmental impacts, waste disposal, decommissioning, governmental protection against cost overruns, insurance & risk of accidents, and owner's costs.

The pros and cons of the costs and economics of power generation options, utilizing renewable energy resources, will continue to be evaluated differently by individuals with widely varying perceptions, and those in authority will most likely dictate the results.

The cost of energy generated by wind power plants has been going down and is now looming to become cost competitive with power generated by conventional energy resources. Costs of wind turbines have been decreasing since the late 1980s, mainly due to:

> ➤ Advancements in technology.
> ➤ Experience gained by operating wind farms.
> ➤ Improvements in quality and reliability of turbines.
> ➤ Developments in building larger multi-megawatt wind turbines.

9.2   COSTS OF LARGE MW WIND TURBINES

The installed costs for a utility scale large MW commercial wind turbine vary depending on the nameplate capacity, the wind turbine, size of the wind farm, and the country where the wind farm is located.

Total installed costs range from about US$ 1.5 million to US$ 2.5 million per MW. At present, most of the commercial-scale turbines installed are about 2 to 3 MW in electrical power rating. Total costs for installing a commercial-scale wind turbine will vary significantly depending on:

> ➤ The number of turbines ordered.
> ➤ Cost of financing.
> ➤ Construction contracts.
> ➤ The location of the project.
> ➤ Several other project cost related factors.

Other factors that will impact project economics include cost components for wind projects such as:

> ➤ Wind resource assessment.
> ➤ Site analysis and Construction expenses.
> ➤ Permitting and interconnection studies.

> ➢   Utility system upgrades.
> ➢   Operations.
> ➢   Warranty and Insurance.
> ➢   Maintenance and Repair.
> ➢   Legal and consultation fees.
> ➢   Taxes and incentives.

The following Figure 9-1 indicates that the cost of energy from renewable and especially wind and solar power is rapidly improving. In the meantime, costs for all non-renewable power plants have substantially increased.

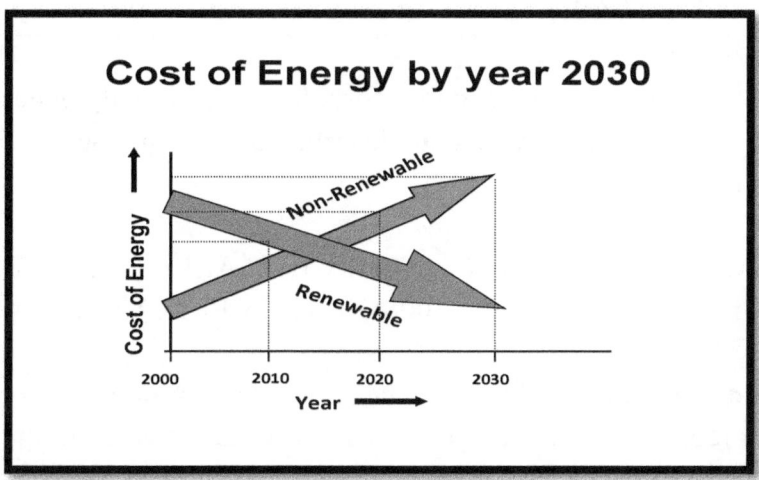

Fig. 9.1 Cost of energy by the year 2030

Due to economy of scale based on mass production, substantial improvement has been achieved in lowering the cost of large MW wind turbines.

In the late 1970s and early 1980s, advancements in technology with the support of government sponsored research and development programs, accelerated the achievement of a substantial reduction in the cost of MW wind turbines to make the cost competitive with conventional power plants using non-renewable energy resources as fuels.

In the year 2016, the costs for MW wind turbines vary, depending on the country of origin, as follows:

> ➢   MW Wind turbine cost =   600 ~ 900 $/kW
> ➢   Total installed cost in a wind farm = 1,000 ~ 2,500 $/kW.

9.3   CONVENTIONAL "COST OF ENERGY" (COE)

At present, the conventional costs for the electricity generated by a power plant is, in general, determined in accordance with the formula:

$$\text{COE (c/kWh)} = \{(CC \times FCR) \times 100 / AE\} + (O\&M \times LF) \times 100/AE + (FC \times LF) \times 100/AE$$

Annual Energy (AE):   The annual energy production for a plant represents the average annual kilowatts delivered to the bus bar during the lifetime of the plant. It takes into account the plant availability and the plant capacity factor.

Capital Cost (CC):   The capital cost of a plant represents the installed cost of the project The fixed charge is the capital annualizing factor which is a function of financial parameters and accounts for the design life of the plant, depreciation, taxes, return to investors, insurance and inflation.

Annual Operation and Maintenance Cost (O&M):   The annual O&M costs are the sum of the total operation and maintenance costs for the life of the project, scheduled and unscheduled costs for the project including inspections, parts and materials, and labor costs. The operation and maintenance costs vary with time due to inflation. Therefore, the O&M costs are leveled by a factor called the leveling factor.

Leveling Factor (LF):   The leveling factor accounts for items such as inflation, which occur during the lifetime of the plant.

Fuel Cost (FC):   The fuel costs for a plant are computed in the same manner as the O&M costs.

The fuel costs vary with locations and fuel pricing policies of the nation. Plant heat rate, cost of fuel and calorific value of the fuel are taken into account in determining the fuel costs.

In the 1960s, conventional fuel costs were low. In the 1970s, their costs increased with higher demand. As the demand for conventional power generation increases, higher costs for the conventional fuels will follow.

Fix Charge Rate (FCR): It is expressed as a fraction of the total installed cost, and it is called the "Fixed Charge Rate (FCR). The fixed charge is the capital-annualizing factor, which is a function of financial parameters, and accounts for the design life of the plant, depreciation, taxes, return to investors, insurance and inflation.

Table 9-1 Fixed charge rate

Fixed Charge Rate			
Life (years)	**Interest Rate [%]**		
	5	7.5	10
10	0.130	0.146	0.163
15	0.0963	0.113	0.132
20	0.0802	0.0981	0.118
25	0.0710	0.0897	0.110
30	0.0650	0.0847	0.106

The fixed charge rate is a leveling factor for capital cost. "Fixed Charge Rate" (FCR) is a function of financial parameters and accounts for items such as the design life of the plant, depreciation, taxes, return to investors, insurance and inflation.

9.4   HIDDEN COSTS

The real constraints on the selection of the energy resources for electrical power generation lie not in the availability of acceptable options. It rests in the political will of utilities and governments to fairly assess the options with a view to providing a risk-free and environmentally acceptable solution. The following Table 9-2 shows the hidden costs of nuclear power energy.

When a fair evaluation is made in comparing the real costs, skeptics always question if the goal of replacing non-renewable energy resources is ever achievable. Creative accounting conventions are often used to justify and to ensure that electrical power production by non-renewable energy resources is cheaper than all other options available on the market.

At present, in accordance with current practices in accounting conventions and economic evaluation of competing energy options for electrical power generation, certain cost items are either deliberately excluded or taken into account by assigning a token value to them to justify calculations.

All such costs that are not fully accounted for in the calculations are "Hidden Costs" that are passed on to the present day taxpayer, or to the future generation of taxpayers. Examples of hidden costs include, but are not limited to, "Decommissioning Costs", "Waste Disposal Costs", "Security of Supply Costs", "Balance of Payment Costs", and "Environmental and Social Impact".

Table 9-2 Hidden costs of nuclear energy

Description	Nuclear Power "Hidden Costs"	Comments – Who Pays? (Government + Tax Payers)
R & D - (1950`1990)	~ 97 billion $	Government + Today's Tax Payers
Insurance	Not Insured –Due to high potential costs	Government + Today's Tax Payers
Cost Overruns	~ 300 % to 700 % +	Government + Today's Tax Payers
Risks – Accident or Terrorist Attack	Single Accident- Two trillion $ +[44000 short term fatalities] [500000= long term fatalities]	Government + Today's Tax Payers
Decommissioning	1 to 2 b$/unit & 10 years	Passed on to Future Generation
Existing Waste Disposal	~ 100 billion $	Passed on to Future Generation
Environmental & Social Costs	~ 3 to 6 c/kWh	Passed on to Future Generation
Government Subsidies	"Hidden" Costs"	Passed on to Future Generation

Decommissioning Costs: The ideal ultimate objective, for the protection of future generations, would be the complete removal and disposal in a manner that the site is rendered safe for habitation. Decommissioning a power plant requires that the plant facilities be permanently withdrawn from operational service and the plant site facilities be put into a safe condition to protect life and the environment. Decommissioning coal-fired plants is relatively simple and cheaper than decommissioning a nuclear plant facility due to the inherent problems associated with the radioactive contamination of the nuclear facilities. To date, there are many nuclear power plants around the world, providing about 15 percent of the entire world's electricity. In due course, these plants will require decommissioning. Decommissioning problems and related costs have not been a serious consideration in the past for any of the conventional plants, because the accounting conventions adopted in evaluating the competing options have ignored the problems.

Security of Supply Costs: Nations that import fuel supplies and/or rely on other nations for the supply of replacement parts and technical staff to help operate the power plants, run the risk of security of supply. Availability of energy resources for production of electrical power is vital to the economic health of any nation. Even a minor interruption in its supply can cause a major disruption in a nation's social and economic life. Utilizing the available resources within national boundaries should be taken into consideration in comparing competing options for power generation.

Waste Disposal Costs: There is a cost associated with the disposal of these waste products. The more stringent the requirements are for safe and risk-free disposal, the more costly the operation. All conventional fuel-burning plants generate atmospheric waste caused by three types of waste, namely - Exhaust gas emissions, Liquid waste, and Solid waste. In an oil burning plant, the problems are disposal of atmospheric and liquid wastes. A coal burning plant faces the problem of disposal of all three types of waste. In a nuclear plant the major problem areas are the disposal of liquid and solid waste. Of all conventional fuel plants, the nuclear plant faces the most serious problems of waste disposal, due to radioactive waste. The disposal of the total nuclear waste accumulated in the USA, from all sources including the military, medical and power plant reactors, since the inception of nuclear power and up to the year 1990, is estimated to be in the order of over 80 billion dollars. In the USA, "about 1,700 tons of highly radioactive waste accumulates each year at power stations". The cost for the partial decommissioning of the 185 MW Yankee Atomic plant in Massachusetts, the first commercial nuclear plant to be decommissioned in the USA, is estimated at US $1,300/kW. This includes the cost for temporary storage of the spent fuel, because no safe permanent storage sites exist at present. The costs for removal to a future disposal site and the costs for the permanent storage facility are not taken into account in the current estimate. The unknown costs for the final disposal of waste from conventional plants are currently not accounted for. Therefore, we have no choice but to pass on the burden of disposing of the waste, at their expense, to future generations. The benefit to the present day tax payer and the consumer is that we can enjoy the cheap electricity from the conventional power plants today by neglecting to take into account the cost of waste disposal.

Environmental and Social Impact: Whenever damage occurs as a result of building and operating a power project, unpaid environmental costs follow. If these costs are not accounted for, they are identified as "Externalities". If the costs for externalities are excluded from the economic analysis process, while comparing the energy options, the results are biased in favor of resources that cause damage to the environment and society.

While there are no agreed methods for calculating or implementing costs for externalities, governments in developed nations, based on research into environmental damages, have taken steps to recognize the problems.

Government regulatory bodies are trying to provide measures for protection of the environment and assign monetary values to the costs of externalities, based mainly on judgment.

At present, only minimum costs are assigned according to what the market can bear, rather than the full cost of the impact of the environmental damage.

As long as the costs for damage to the environment are not taken into account during the economic assessment of the alternatives for power generation, safe and risk-free energy sources will be competing against energy options presently under the protective umbrella of "Hidden Costs".

Balance of Payment Costs: When the fuel for power generation has to be imported it imposes a drain on the country's economy. Utilizing resources within national boundaries and reducing the need for foreign fuel imports can reduce the money drain and the trade deficit.

9.5   LIFE CYCLE ASSESSMENT (LCA)

Life Cycle Assessment (LCA): LCA is one of the tools that could be utilized to assess the sustainability of renewable energy sources. Life Cycle Assessment is an approach in which all energy required to make a product are accounted during its full three phase life cycle, such as:

> ➢   Phase 1 - resource extraction.
> ➢   Phase 2 – manufacturing.
> ➢   Phase 3 - final disposal.

Table 9-3 indicates a comparison of "Net Energy Ratio" to determine the true value of energy of selected technologies.

Table 9-3 Life-cycle assessments of selected technologies

Fuel	Strong Return on Energy Investment
	Net Energy Ratio
Hydro	30.9
Wind	47.4
Solar	3.62
Coal	0.341
Nat. Gas	0.40
Nuclear	0.313

LCA is considered as "cradle to grave" analysis. It means taking into account all the energy that is required to obtain the fuel and utilize it to generate electricity. Strong return on energy investment is indicated by net energy ratio values greater than 10.

Therefore, arguments and advantages to switch over to using renewable energy resources become undeniable.

9.6   VALUE OF ENERGY OPTIONS

Liberties taken with Figures while assigning values to the costs of various components in the formula for computing the cost of energy, results in a bias for the pre-selected and preferred option.

One of the methods commonly in use is to omit certain costs for the preferred option. Examples of such practices are seen in calculations that omit certain costs such as: cost of actual funding required during construction, decommissioning and waste disposal, etc. Another method, which is very prevalent in the accounting convention, is to bend the facts so that the costs of the preferred option get the desired end-results.

By applying values to lower the costs for items such as capital, operation and maintenance, fuel, life of project, capacity factor, inflation and fixed charge rates, the figures can be made to favor the preferred option.

A review of past studies and reports indicates that both of these methods are routinely practiced in the power generation industry. In making comparisons for assessing the economic viability of energy options for electrical power generation, the value assigned to items such as, "Capacity Factor", "Fixed

Charge Rate", "Project Size", and "Life of the Project", play an important and major role in the final outcome of the results.

Capacity Factor: One of the biggest problems in projections of power generation costs is one of selecting a reasonable and justifiable capacity factor for the calculation of COE.

In the case of coal and nuclear plants, the value assigned to capacity factor plays a major role since the high cost of the unit has to be amortized over the maximum possible number of kilowatts.

Depending on the assumptions made for the overall capacity factor based on the total life of the plant, the projected costs can be made to go in any direction. Using high values for capacity factor say 80% or 90% for a coal-fired or say 90% or 95% for a nuclear power plant would give biased results.

Fixed Charge Rate: One of the major factors between the differences in COE, for a plant financed by a utility and the COE for a plant financed by an independent power producer, lies in the ability of utilities to obtain project funding at a reasonable rate. Utilities can raise money for capital costs at far lower rates than small independent power producers. Currently, the fixed charge rate used in calculating COE varies, from about 3 % to 5 % for utilities to a high of 7 % to 10 % for an independent power producer.

Project Size: Selection of similar plant size is important to the outcome of the result of economic evaluation. For example, comparison of COE for two nuclear plants of similar design but different plant size, one-500 MW unit and the other at 1000 MW unit, would be biased in favor of the larger unit because of the economy of scale. Similarly, comparison between a 750 MW coal-fired plant with a 500 MW nuclear plant would favor the coal fired plant. Utilities evaluating wind power should give due consideration to the size of the wind farm. Evaluating a 3, 5 or 10 MW wind farm against COE of generating power from a large coal or nuclear plant cannot be justified as a fair comparison. Economy of scale is just as valid for wind farms as it is for conventional plants.

Life of Project: All COE comparisons for different options should be based on the same duration for the operating life of the project. Results of economic evaluation can be tilted in the favor of a plant by allowing say a 40 to 50-year life as compared to allowing a plant a 20 or 30-year life.

Other Credits: Other credits are of interest to governments only. These can include balance of payment credit for reducing the import of conventional fuels, social cost credits and environmental cost credits. The full value of social and environmental credits is, at the best of times, very difficult to measure. This would depend on the type of fuel displaced and the value society attaches to a clean environment and risk-free energy option.

Value of kWh: A   kWh   of   electrical   power production requires about 10 mega joules in the form of conventional energy resource. The value of a net kWh produced from a renewable resource such as wind represents the cost of 10 mega-joules of conventional fuel, which otherwise would have been utilized to generate the electricity.

From an economic point of view, it is important to take into account the type of fuel most likely to be saved due to differences in the costs associated with the conventional fuels. Energy from wind is intermittent, by its inherent nature. Therefore, to utilities, the value of the electrical energy (kWh) generated from wind turbine power plants depends upon the types of conventional generating plants in the system, system loads, reliability requirements, fuel costs and other financial parameters.

Because of the intermittent nature of the wind, only about 20 to 30 percent maximum penetration in the power systems is currently regarded as economically viable. Therefore, utilization of wind power, without any additional energy storage, does not in any way threaten the use of conventional options such as coal or nuclear.

Furthermore, the value of energy from wind depends upon different factors such as savings of fuel, the reduced capital investment required for new conventional equipment in the expansion program and base load capacity credit.

If a utility has storage capacity using either pumped storage, hydroelectric, or in the future some storage system such as compressed air, hydrogen, or batteries, it can utilize wind power beneficially.

In such cases wind power should receive capacity credit. It would also increase the percentage of maximum penetration in the power systems, from 20 % to 30 % that is currently regarded as economically viable, to a much higher level, depending on the utility's storage capacity.

At present, utilities with hydroelectric plants and wind farms in their system mix can offer the maximum capacity credit, because the required storage capacity is an inherent characteristic of the hydro plants. It is paid for, and it is there ready to be utilized.

Also, for about one-third of the cost of a total full-scale project, an additional hydro power plant can be installed at the existing plant site without the necessity to increase the water reservoir capacity.

This combination, of wind power and additional hydro plant without the necessity to increase the water reservoir, provides additional firm power generation capacity in the system. In such cases wind power should receive full capacity credit, since the combination of storage system and the wind power represents dependable capacity.

9.7   REAL COST OF ELECTRICITY

To arrive at the real cost of electricity to the consumer and taxpayers, it is essential to understand costs that influence the production of electricity for a given power generating facility, be it by conventional non-renewable or renewable energy resources. The solution to the problems related to the comparative "Costs and Economics of Wind Turbines" lies in:

➢ Eliminating all energy subsidies, both direct and indirect, or subsidizing them equally to level out the playing field.

➢ Evaluating the comparative cost of energy (c/kWh) impartially, and taking into account all costs, including miscellaneous costs, such as: both direct and indirect costs, governmental protection against cost overruns, insurance and risk of accidents costs, owner's costs, and balance of payment costs, including miscellaneous (hidden) costs, such as:

> ➢   Social Impact.
> ➢   Plant Security.
> ➢   Environmental impacts.
> ➢   Waste disposal.
> ➢   Decommissioning.

Economic considerations underlie nearly all-major decisions in selecting the deployment of energy options for electrical power generation.

The following description provides a brief explanation of how the cost of energy is established, what are the factors used for comparing the cost of producing electricity?

What is the true cost of producing electricity? The major items of costs that influence the "Cost of Energy" (COE) of electrical power generation are:

> ➢   Hidden costs.
> ➢   Front-end costs.
> ➢   Funding for R&D.
> ➢   Government subsidies.
> ➢   Operation and maintenance costs.
> ➢   Fuel costs.

Hidden Costs: Plant   decommissioning   costs,   waste disposal costs, security of supply costs, social impact costs, and environmental impact costs.

Front-End Costs: Costs such as R&D funding are essential to the development of technologies, for exploiting energy resources for power generation and to provide security of supply for the future.

Assistance provided to energy programs brings about improvements in the existing technologies and the commercialization of new products. R&D funding indicates that of the billions of dollars requested for energy R&D programs, a majority of the funds are allocated to nuclear related projects, clean coal technology and fossil energy R&D, and less than 5% to renewable energy R&D. This type of disparity in R&D funding is one of the reasons for the slow growth in the development of renewable energy options for power generation.

Government Subsidies: Fossil fuels such as coal and oil receive direct subsidies, and indirect subsidies in the form of taxpayer support for external costs such as disability payments for coal miners, military expenditures to protect oilfields, and environmental pollution which intensifies health care costs.

Operation and Maintenance Costs (O&M): O&M costs for both conventional power plants using non-renewable fuels and power plants using renewable fuels are similar. Improvements in design and technology of wind and solar power plants indicate that, in the future, O&M costs for wind and solar plants will decrease.

Fuel Costs: Power plants using non-renewable energy resources of fuels such as coal, oil, natural gas, and nuclear fuels have associated costs, depending on the type of fuel used to generate power. Power plants using renewable energy resources such as wind and solar have zero fuel costs.

9.8   COMPARISON OF REAL COST OF ENERGY (COE)

The following Table 9-4 provides the comparative cost of electricity. It illustrates that the electrical power generated by wind power can be cost comparative with electricity produced by coal and nuclear power plants.

Table 9-4 Comparative cost of electricity

Comparative Cost of Electricity

(Average costs in US cents / kWh – 2016)

Description	Coal Plant	Nuclear Plant	Wind Plant
Capacity (MW)	1000	1000	3000
Capital Cost ($ B)	2 – 8	4 - 8	4.5 – 7.5
Energy (GWh/yr)	7884	7884	7884
Plant Cost (c/kWh)	1.65 – 6.6	3.3 – 6.6	3.7 – 6.23
Plant O&M (c/kWh)	0.5 ~ 1.0	1 ~ 2	0.5 ~ 1
Fuel Cost (c/kWh)	1.5 ~ 2.0	0.5 ~ 0.6	0 ~ 0
COE (c/kWh)	3.65 ~ 9.6	4.8 ~ 9.2	4.2 ~ 7.23
Hidden Costs* (c/kWh)	4.56 ~ 7.12	4.28 ~ 10.64	0.22 ~ 0,49
Real Cost	8.21~16.72	9.08~19.84	4.42 ~ 7.72

{*Hidden Costs = Government Subsidies; Protection against Cost Overruns, Insurance & Risk of Accidents; Environmental Impact; Waste Disposal; Decommissioning; Owner's Costs }

9.9 SUMMARY

✓ The electrical power rating of most of the large wind turbines installed today in wind farms is about 2 to 3 MW. Depending on the location, country, and the size of the wind farm, the installed costs for a utility scale large MW wind turbine range from about US$ 1.5 million to US$ 2.5 million per MW of nameplate capacity.

✓ As shown in the Table 9-4 when comparison is based on "Real Cost", wind energy is cost competitive with conventional non-renewable energy resources.

✓ "Creative Accounting" conventions are often used to ensure the selection of an energy resource option, which could not otherwise be justified. All such costs that are not fully accounted for in the calculations are " Hidden Costs".

✓ When a fair evaluation is made in comparing the available options for power generation, then, in regions with good wind resources, wind power can compete with available conventional power generation for the production of electricity.

✓ All nations must utilize all available energy resources including oil, coal, natural gas, nuclear, hydro, wind and solar in a safe and clean manner, to provide maximum protection of the consumers, without passing the burden f cleanup and waste disposal as our legacy to future generations.

CHAPTER   10

FUTURE

OF THE

WIND TURBINE

INDUSTRY

10.1   OVERVIEW

Since the early 1970s, the main focus of the wind turbine industry was to develop and build wind turbines that would generate electricity as cheaply as technologies that utilize fossil fuels - (@ 2 to 4 cents/kWh).

Increased funding for research and development for wind turbines enabled the wind industry to install and test prototypes, and gain valuable experience in the operation and maintenance of wind turbines.

This resulted in the development of present-day designs of both the HAWT and the VAWT.

The designs of present-day wind turbines have improved significantly, resulting in an increase in reliability and affordability.

In regions with good wind conditions (class 3+), wind turbines are cost competitive with conventional non-renewable energy options such as coal, oil, and nuclear.

Wind energy is a source of energy that is everlasting. In many areas of the world, with adequate wind energy resources, wind power is rapidly becoming a viable energy option. Utilizing wind energy for the generation of safe and clean power will provide maximum protection to consumers, taxpayers, and ratepayers.

Hybrid wind/solar systems will play a major role in supplying electrical power needs for all sectors of the economy.

Problem: The future of the wind turbine industry depends on:

> ➤ Eliminating the inherent problems related to designs of existing wind turbines on the market.
> ➤ Designing wind turbines with innovations that specifically address clients' concerns and demands with respect to acceptability, reliability, and affordability.

Solution: Designs of existing wind turbines must be improved, to make the wind turbines acceptable, reliable, and affordable.

This can only be achieved by eliminating the inherent limitations associated with their designs. Advancements in wind turbine technology now make it possible to eliminate the inherent design limitations in existing wind turbines on the market.

Future: The future of the wind turbine industry will be governed by how the wind turbine industry adapts to innovations in the designs of wind turbines that specifically address the criteria for public concerns with respect to acceptability, reliability, and affordability.

Harnessing the wind's energy with wind turbines designed with the latest advancements in wind power technology would make wind power acceptable, reliable, and affordable. It would:

✓   Eliminate NIMBY syndrome
✓   Be cost competitive with other energy resources
✓   Enhance market growth
✓   Improve sales

10.2 PROBLEMS AND SOLUTIONS

Design criteria for public acceptance is outlined in the following Table 10-1.

Table 10-1 Design criteria for public acceptance

(Acceptable; Reliable; Affordable)		
Acceptable	**Reliable**	**Affordable**
✓ Noise free operation ✓ Safe in operation ✓ No birds killed ✓ 100% lightning protection ✓ No restriction on location ✓ Meet all local by-laws ✓ Aesthetically pleasing design ✓ Low visual impact	✓ Rugged design and construction ✓ Safety in Operation ✓ Longer service life (>30 years+) ✓ Less maintenance	✓ Maximum Annual Energy production ✓ Competitive Energy Costs (c/kWh) ✓ Higher Internal Rate of Return ✓ Lower Payback Period

Problems: Based on the operational experience to date, the main areas of problems with the existing designs of both the Small and the Large MW wind turbines are: reliability, acceptability, and affordability. Failure to meet criteria for public acceptance is the direct consequence of the public's opposition to installations of wind turbines in their neighborhoods. This results in a symptom that is commonly known as "NIMBY" (Not In My Back Yard) syndrome.

For these reasons the existing wind turbine industry is:

➢ Unable to meet the existing demand, and
➢ Powerless to capture the growing market for wind turbines.

Solutions: At present, there is a wide variety of Research and Development (R&D) work being carried out to bring about advancements in the technology and design of shrouded wind turbines. The aim of the advancements is to make wind turbines acceptable, reliable, and affordable, with a goal to:

➢ Meet the existing demand.
➢ Capture the growing market for wind turbines.

10.3   REVIEW OF WIND TURBINE DESIGNS

Wind energy, in the early 1970s, was considered impossible as a viable energy option, but now it is one of the leading energy options competing with non-renewable conventional energy options.

Advancements in shrouded wind turbine technology now make it possible to design wind turbines that are: acceptable, reliable, and affordable.

A review of the past historical records of the development of wind turbines indicates that utilization of the principle of augmentation of wind with structural design features dates back to the early windmills in Persia and China.

Table 10-2 Comparative design analysis

#	Criteria for Public Acceptance	Past		Present		Future	
		Yes	No	Yes	No	Yes	No
Acceptable							
1	Noise free operation	✓			X	✓	
2	Safe in operation	✓			X	✓	
3	No birds/bats killed	✓			X	✓	
4	100% lightning protection		X		X	✓	
5	No restriction on location	✓			X	✓	
6	Meet all local by-laws	✓			X	✓	
7	Aesthetically pleasing design	✓			X	✓	
8	Low visual impact	✓			X	✓	
Reliable							
9	Design & construction	✓		✓		✓	
10	Safety in operation	✓			X	✓	
11	Longer service life	✓			X	✓	
12	Less maintenance	✓			X	✓	
Affordable							
13	Maximum annual energy production	✓			X	✓	
14	Comparative energy costs (c/kWh)	✓		✓		✓	
15	Higher internal rate of return	NA			X	✓	
16	Lower payback period				X	✓	

The comparative analysis of wind turbine designs indicates the inherent design problems associated with present-day wind turbines.

Technological advancements in design indicate that wind turbines with a shrouded design will have a significant impact in overcoming the "NIMBY" syndrome.

Wind power production is proportional to the wind speed cubed. Therefore, for any increase in the flow and velocity of the wind entering the turbine, there is a large increase in power output. It will produce higher annual energy output, even in areas where lower wind speeds and complex wind patterns are expected.

10.4   WIND TURBINES WITH SHROUDS

Through state-of-the-art designs and the advanced technologies of shrouded wind turbines, it is now possible to produce more annual energy than present day traditional wind turbines. (Depending on the design, about 50 to 150% more energy than the present day wind turbines without shrouds).

Numerous research studies have been carried out related to the design and engineering of wind turbines with shrouds. For both types of wind turbines, HAWT and VAWT; prototype projects have been built, installed and tested.

At present, there are many designs of HAWT and VAWT units with shrouds that are in the developmental stage and most of them have not yet reached commercialization.

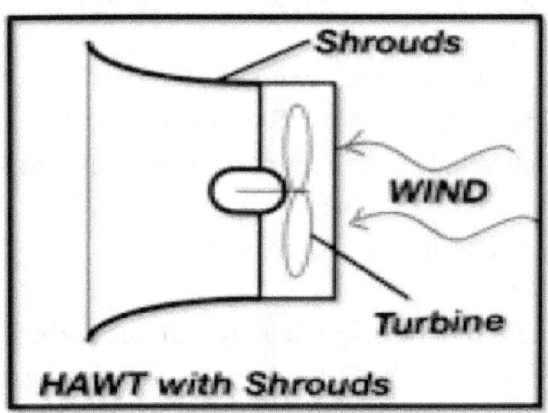

Fig. 10-1 Design of small HAWT with shrouds

The following illustration shows the advanced design features and the advantages of the VAWT Wind Turbine with shrouds.

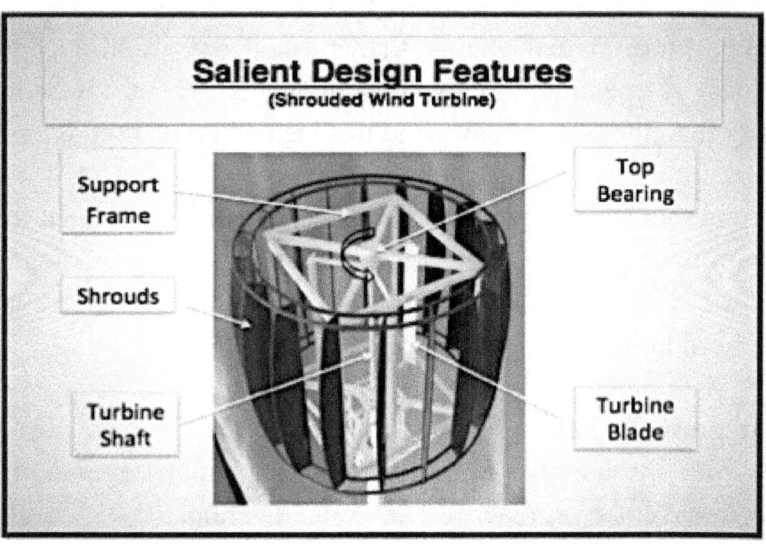

Fig. 10-2 Design of small VAWT with shrouds

What distinguishes the wind turbine is its patent and innovations in design features. Shrouded wind turbines, with a patented straight-bladed vertical axis, as shown, are specifically designed to address clients' concerns with respect to acceptability, reliability, and affordability.

Each of the salient design features of the turbines designed with shrouds serves a specific purpose. Enclosing the turbine with a support frame provides a rugged support system. In addition, it provides total lightning protection. The shrouds improve aesthetics, and increase the power output of the turbine. Shrouded design overcomes "NIMBY" syndrome related to wind turbines.

Wind turbines with designs incorporating shrouds help in achieving the goal of making them acceptable, reliable, and affordable.

The following Figure 10-3 illustrates the advanced design features of a straight-bladed small VAWT unit with shrouds.

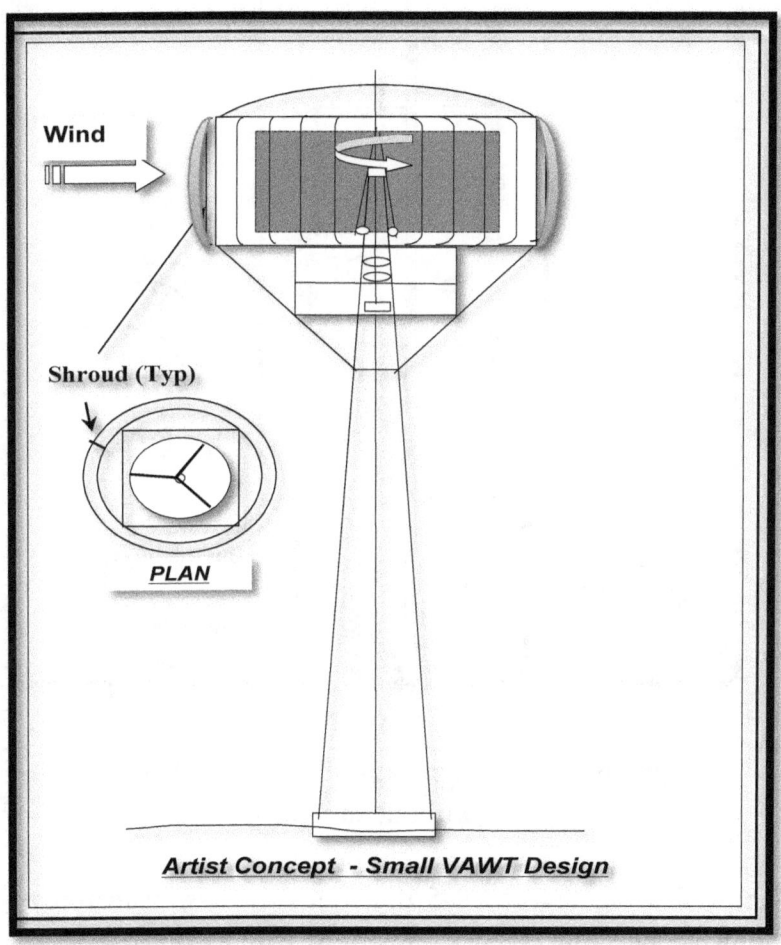

Fig. 10-3   Design concept of a small VAWT with shrouds

The following Figure 10-4 illustrates the advanced design features of a straight-bladed large MW VAWT unit without shrouds.

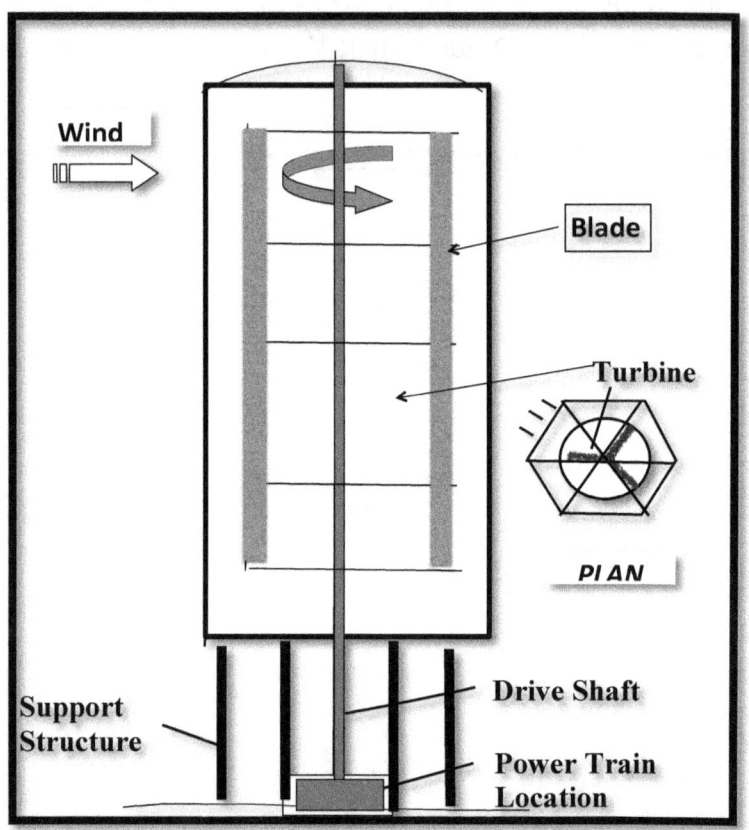

Fig. 10-4 Design concept of a large MW VAWT
without shrouds

The following Figure 10-5 illustrates the design features of a straight-bladed large MW VAWT unit with shrouds. The unit is specifically designed to address clients' concerns for acceptability, reliability, and affordability.

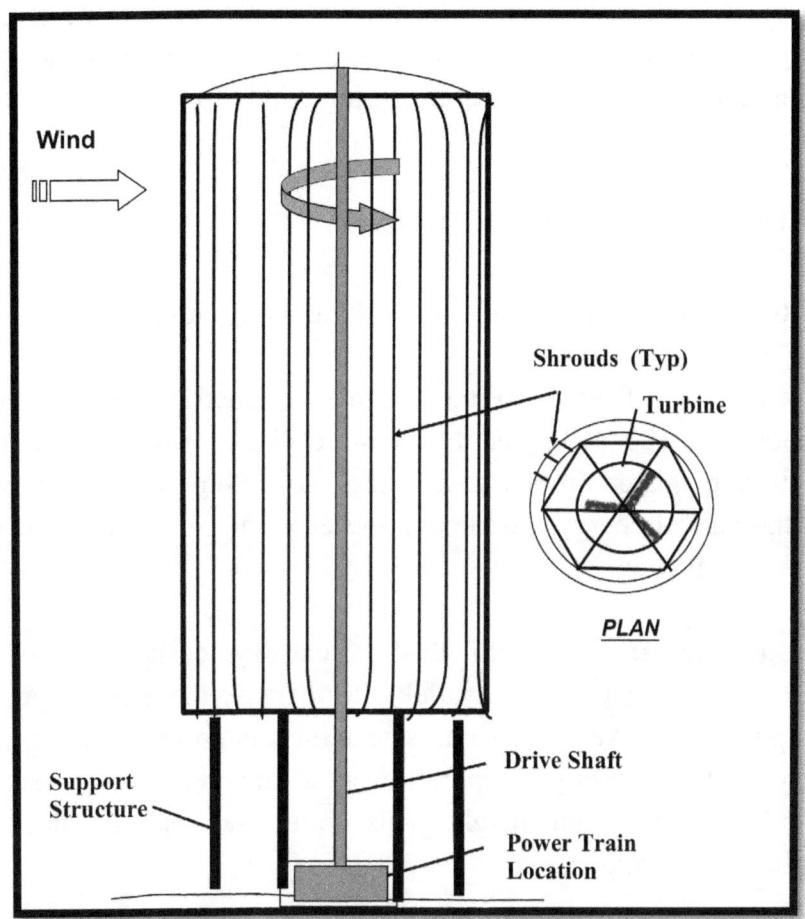

Fig. 10-5 Design concept - MW VAWT with shrouds

Wind turbines with shrouds represent a "state-of-the-art" design that substantially increases annual energy output. Hence, it makes it possible to harvest wind energy potential in areas with low wind conditions. (Class 2).

Wind turbines with shrouds resolve the public's concerns with respect to "NIMBY" syndrome, which leads to making wind turbines "acceptable".

With a clearly superior product, the shrouded vertical axis wind turbines will be well placed to compete with local and international companies manufacturing wind turbines.

10.5   COSTS AND ECONOMICS

Costs and economics of wind turbine units with shrouds are:

Costs: Costs of Wind Turbines designed with shrouds are about 15 to 20 % higher than units designed without shrouds. However, this additional cost is easily overcome by an increase in the annual energy production and economy of scale based on mass production.

Economics: With the present-day designs of wind turbines, the only option available to increase the annual energy output is to make the turbine size larger and/or to increase the height of the turbine support tower, to achieve an increase in wind velocity at hub height. This option increases the cost of turbine installation.

Wind turbines designed with shrouds help to:

> ➢   Increase amount of wind flow to the turbine.
> ➢   Increase speed of wind flow to the turbine.

The following Table 10-3 shows the comparison of annual energy production of a vertical axis wind turbine designed without and with shrouds.

Table 10-3   Comparison of VAWT design

Performance: VAWT Without Shrouds	
Wind Speed at Site m/s (mph) @ Hub Height	Annual Energy – kWh / year
4.0 (8.95)	5,000

Performance: VAWT With Shrouds				
Wind Speed at Site m/s (mph) @ Hub Height	Annual Energy kWh / year Without Shrouds	Wind Speed at Turbine m/s (mph) With Shrouds	Annual Energy kWh / year With Shrouds	Increase in Annual Energy % With hrouds
4.0 (8.95)	5,000	4.5 (10.07)	7,500	50
		5 (11.12)	10,000	100
		5.5 (12.3)	12,000	120
		6 (13.42)	15,000	150

Depending on the size and the design of the shrouds, the increase in wind speed would vary. The increase in annual energy production could be in the range of 50% to about 150%. It would improve the economics and help to make wind turbines affordable.

10.6   SUMMARY

Wind Energy: For an on-demand power supply, the basic difference between renewable and non-renewable energy systems is in the location of the energy storage system. Non-renewable energy systems require energy storage before the generation of power, while wind energy systems require energy storage after the generation of power. Hybrid wind/solar systems with an energy storage system will play a major role in supplying electrical power needs for all sectors of the economy.

Wind Turbines – What Works? At present, the designs of both the Horizontal-Axis (HA) and the Vertical-Axis (VA) MW units are reliable and affordable. HAWTs are cost competitive in domestic and international markets for wind farms. HAWTs are currently generating the vast majority of wind power in today's market. The wind power industry has matured around HAWTs.

What Does Not Work?   Existing wind turbines on the market face many obstacles, including the existing State or local municipal zoning laws, which may result in expensive hearings or could possibly prevent installation

Based on the operational experience to date, the main areas of problems with the existing designs of both the Small and the Large MW wind turbines is acceptability. Failure to meet criteria for public acceptance is the direct consequence of the public's opposition to installations of wind turbines in their neighborhoods. This results in a symptom that is commonly known as "NIMBY" (Not In My Back Yard) syndrome.

For these reasons the existing wind turbine industry is:

> ➢ Unable to meet the existing demand, and
> ➢ Powerless to capture the growing market for wind turbines.

And Why? Even though designs of wind turbines show significant improvements, there still remain a number of major issues with existing limitations in the inherent design of the HAWTs.

The major inherent limitations of the design of HAWTs are:

> ➢ Safety in operation
> ➢ Birds and bats killed by the turbines
> ➢ Problem of noise generated by the rotation of the turbine blades
> ➢ Damage and fire caused by lightning strikes on the turbine

Future? Harnessing the wind's energy with wind turbines designed with the latest advancements in wind power technology would make wind power acceptable, reliable, and affordable. It would:

> ✓ Eliminate NIMBY syndrome
> ✓ Be cost competitive with other energy resources
> ✓ Enhance market growth
> ✓ Improve sales

The future of the wind turbine industry depends on how the wind turbine industry adapts to innovations in the designs of wind turbines that specifically address the criteria for public concerns with respect to acceptability.

APPENDIX 1   GLOSSARY

CAPACITY FACTOR (CF):
A measure of power production over a given time period, divided by the amount of annual energy production, running at 100% capacity during that same time period.

CAPITAL COSTS:
The total installed cost for a wind turbine.

ENERGY:
Energy can be converted into different forms - such as kinetic, potential, chemical, and/or electricity. The total amount of energy always stays the same.

EXTERNALITY:
The values that go together with an economic business deal, where it influences others beyond the direct financial performers.

FREQUENCY:
The number of cycles through which an alternating current passes per second, measured in hertz.

FMEA AND FTA:
Failure Mode Effect Analysis, and Fault Tree Analyses.

GEARBOX:
A system of gears used to increase or decrease shaft rotational speed.

GENERATOR:
A machine for converting mechanical energy to electrical energy.

GIGAWATT (GW):
A unit of power - equals to one million kilowatts.

GIGAWATT-HOUR (GWH):
A unit of electricity - one million kilowatts (kW) expanded over a period of one hour.

GLOBAL WARMING:
A term used to describe the increase in average global temperatures caused by the greenhouse effect.

GRID:
A term that signifies an electricity transmission and distribution system. It is also known as utility or a power grid.

KILOWATT (KW):
A unit of electrical power (watt) times 1,000 equals one kilowatt.

KILOWATT-HOUR (KWH):
A measure of 1 kW electricity - expanded over a period of one hour.

MEGAWATT (MW):
A measure of electricity, 1 MW is equal to 1,000 kilowatts.

TERAWATT: (TW):
A measure of electricity, 1 TW is equal to 1,000 megawatts.

RADIOACTIVE WASTE:
Waste residuals produced from generating electricity from nuclear fuel.

R&D:
Research and Development

RENEWABLE ENERGY:
Energy resources that are regenerative or that cannot be
depleted, such as: wind and solar.

ROTOR:
Rotating components of a wind turbine.

SOLAR ENERGY:
Energy conveyed from the sun (solar radiation).

UTILITY GRID:
Term refers to transmission and distribution system of
electricity, by an electrical utility.

WIND TURBINE:
A term used for a device that converts wind energy to
generate electrical power.

WIND POWER:
Power produced by a wind turbine to convert the available
power in the wind into electrical power.

WIND POWER DENSITY:
The wind power density, measured in watts per square meter
(W/m^2), indicates the amount of energy that exists at the site
for conversion to power by a wind turbine.

WIND POWER CLASS:
A scale devised for classifying wind power density. There
are seven (7) wind power classes, 1 being the lowest to 7
being the highest.

APPENDIX 2   BIBLIOGRAPHY

[1]   Canadian Wind Energy Association (CanWEA).] "Wind Energy Basic Information", -

[2]   S. Quraeshi, B.M. Pederson, and Dr. A.A.M. Sayigh, - "Wind Turbine generators: State-of-the-art", Solar and Wind Technology, An International Journal, Vol. 1, 1984 (p 37).

[3]   S. Quraeshi - "Renewable energy – the key to a better future", Solar and Wind Technology, An International Journal, Vol. 1, 1984 (p 25).

[4]   S. Quraeshi - "Large wind turbine generators (WTGs) for electrical utility application". Presented at 11th World Energy Conference, September 1980, Munich, Germany.

[5]   S. Quraeshi and Brian Richards - "Wind Power-A viable energy option", Energex '82 conference, Regina, Saskatchewan, Canada.

[6]   S. Quraeshi and Brian Richards - "Application of Large Wind Turbine Generators in Utility Networks", presented at 6th Annual Wind Energy Symposium, Toronto, Nov. 1983

[7]   S. Quraeshi - "Review of renewable energy technologies applicable to Caribbean", Electrical Utilities Conference, May 27-29, 1981, Barbados, West Indies.

[8]   Shawinigan Report - "Study of Large Wind Turbine Generators for Electrical Power Generation", July 1980.

[9]   S. Quraeshi - "Utility Grade Wind Farms", CanWea Conference, 1993, (p 245).

[10]   S. Quraeshi   -   "Costs and Economics of Power Generation", Canadian Wind Energy Conference, 1992, (p 262)

www.ingramcontent.com/pod-product-compliance
Lightning Source LLC
Chambersburg PA
CBHW071429170526
45165CB00001B/449